Further Supplement Issues of Zeitschrift für Kristallographie

The Mathematics of Structures

The Exponential Scale

Professor Dr. Sten Andersson
and
Dr. Michael Jacob

R. Oldenbourg Verlag München Wien 1997

Address of the authors:

Professor Dr. Sten Andersson
Sandviks Forskningsinstitut,
S Långgatan 27, 380 74 Löttorp, Sweden

Dr. Michael Jacob
Inorganic Chemistry
Arrhenius Laboratory
Stockholm University,
106 91 Stockholm, Sweden
e-mail: jacob@inorg.su.se

© 1997 R. Oldenbourg Verlag
Rosenheimer Straße 145, D-81671 München
Telefon: (089) 45051-0, Internet: http://www.oldenbourg.de

Gedruckt auf säure- und chlorfreiem Papier
Gesamtherstellung: R. Oldenbourg Graphische Betriebe GmbH, München

ISBN 3-486-64258-8
ISBN 978-3-486-64258-2

TABLE OF CONTENTS

INTRODUCTION

We are chemists. Why would we worry about mathematics? We will try to explain.

We are builders. We see that complicated structures can be described with simple building blocks, with units of simple structures[1,2]. We are constructing the mathematics behind them. The words of Hildebrandt, *It is a basic principle of mathematics to build more and more complicated structures out of simple ones,* give great comfort in our search[3].

We are of course seeking different understandings. The chemist talks about the nature of the chemical bond and the physicist talks about the nature of motion when trying to understand interactions, which means forces between bodies. It may be planets, atoms, molecules, electrons or other small particles. For the chemist this is the key to the understanding of reactivity, properties, of a solid or a molecule, to the origin of life. For a physicist it is the key to travels in space, nuclear structures, or another bomb.

For a chemist it is of vital importance to have a developed picture of the nature of bonding. He may be in the trade of solid state, in the trade of drugs, in the trade of enzymes. A chemist knows where the electrons are before he makes the calculation to find them. Pauling did. Roald Hoffmann and von Schnering do. The chemist wants to understand the structure of a new alloy, a ceramic or a superconductor, or of an enzyme or a drug. He wants to understand the **structure,** where the atoms are, and why they are there, and what are the forces holding them together. This means the nature of interaction.

So we shall study structures - from the structures we deduce the rest.

We shall soon describe some mathematics which we found extremely useful and call **The Exponential Scale.** But first we need to describe some fundamental mathematics and chemistry. We must presume that you understand some mathematics as we give some of the jargon used in the differential geometry needed.

Dedicated to Hans Georg von Schnering

CURVATURE

Take a surface and let a plane rotate through a surface point in its normal. The section of this normal plane and the surface is a curve of curvature k. During the rotation, k must attain one maximum and one minimum value, k_1 and k_2. They are called principal curvatures, and corresponding planar curves principal lines of curvature. These two curvatures are very useful in the description of the properties of surfaces. Their product is the *Gaussian curvature* (K)

$$k_1 k_2 = K \qquad\qquad (1)$$

and the *mean curvature* (H) describes the sum

$$k_1 + k_2 = 2H. \qquad\qquad (2)$$

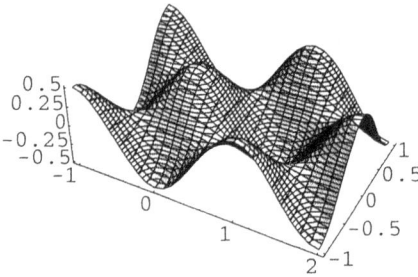

Fig 1. A doubly periodic surface. Positive Gaussian curvature in the peaks and negative in the valleys, or saddles.

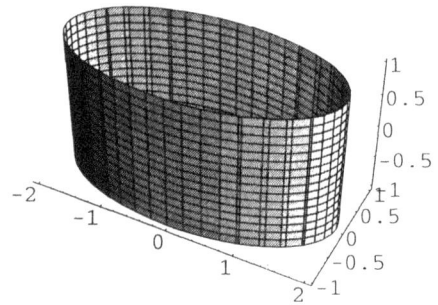

Fig 2. Elliptic cylinder with zero Gaussian curvature. Points on it are called parabolic.

A typical surface is shown in fig 1. It contains positive as well as negative Gaussian curvature, and it is doubly periodic as is obvious from its equation:

$$\cos(\pi x)\cos(\pi y) - 2z = 0 \qquad\qquad (3)$$

We will now show some important examples of surfaces and discuss their curvatures.

A plane has both K=0 and H=0, while a cylinder has K=0 for every point as one of the principal curvatures is a straight line.

The mean curvature is 1/r for an ordinary cylinder. The surface in fig 2 is an elliptic cylinder with the equation

$$x^2 + 4y^2 = 4 \tag{4}$$

A point on a cylinder is called parabolic.

A point on a sphere or an ellipsoid has always positive curvature and for the sphere $K=1/r^2$ and $H=1/r$. Such a point is called elliptic. In figs 3 and 4 we see a sphere and an ellipsoid with the equations

$$x^2 + y^2 + z^2 = 4 \tag{5}$$

$$2x^2 + y^2 + z^2 = 4 \tag{6}$$

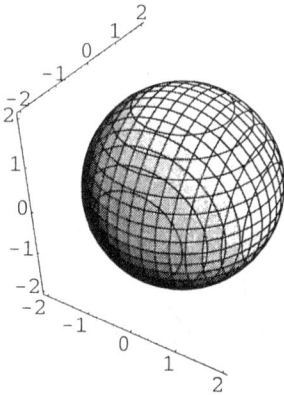

Fig 3. Sphere. Positive Gaussian curvature.

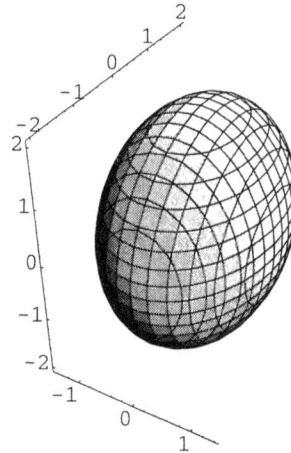

Fig 4. Ellipsoid.

A simple example of a surface of negative Gaussian curvature is the saddle. An example of this is shown in fig 5, with the equation

$$x^2 - y^2 - z = 0 \tag{7}$$

A point on such a surface is called hyperbolic.

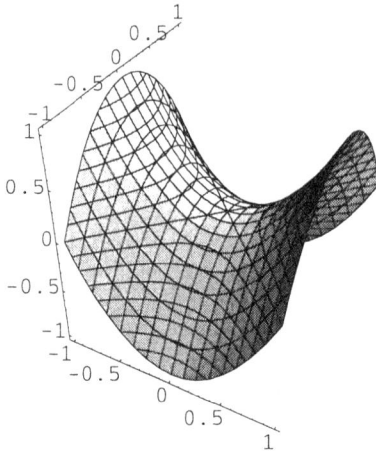

Fig 5a. Saddle. Negative Gaussian curvature, points are called hyperbolic.

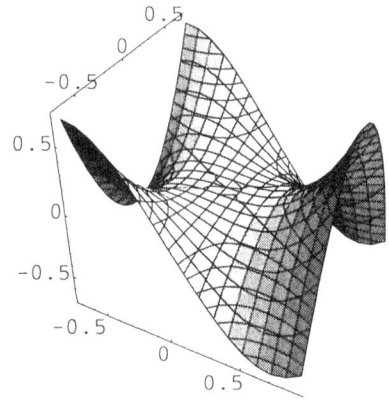

Fig 5b. Monkey saddle with a flat or umbilic point. Negative Gaussian curvature everywhere else.

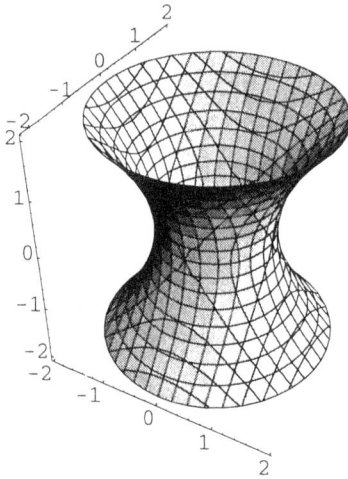

Fig 6. The catenoid.

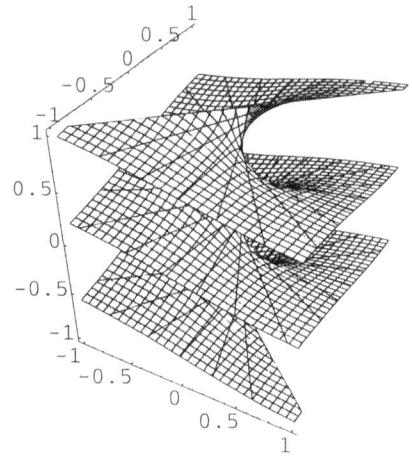

Fig 7. The helicoid.

Examples of surfaces built of saddles are the catenoid and the helicoid with the equations

$$x^2 + y^2 - \cosh z = 0 \text{ (catenoid)} \qquad (8)$$

$$x\cos 4z - y\sin 4z = 0 \text{ (helicoid)} \qquad (9)$$

Both are minimal surfaces, or soap water surfaces. Another way saying this is that H=0. These two surfaces are very special, they have the same Gaussian curvature on corresponding points. That means they are isometric and can be bent into each other without stretching, like a paper can be rolled into a cylinder. This has been used by us to describe phase transitions without cost of energy in liquid crystals and martensite[12,13]. It is called the Bonnet transition.

The monkey saddle in fig 5b is a very remarkable surface. Hilbert gave it the name - a monkey beside its two legs also has a tail. The monkey saddle has negative Gaussian curvature everywhere except in the centre where it is zero. Such a point is called umbilic or a flat point. We have used a simple function (10) to show a monkey saddle in 5b.

$$x(x^2-3y^2) - z = 0 \qquad (10)$$

Of special interest in mathematics are functions of constant Gaussian curvature. The simplest for the positive case is clearly the sphere. For constant negative Gaussian curvature there are several special cases as recently summarised[4]. The general case is beautifully described by Thurston[5]. A structure of such a geometry is truly isotropic - wherever you are situated, it is the same, and in whatever direction you look, you see the same thing. Consequently we have developed these thoughts to describe the structure of water or glass[6].

Monkey saddles and ordinary saddles build the strongly related and famous 3- periodic nodal and minimal surfaces. Some part of this book deals with that type of surfaces. The minimal surfaces are well characterised, having H=0 everywhere and K≤0. We show the nodal surface (equation 11), which deviates within 0.5% from the P - minimal surface, in fig 8.

$$\cos x + \cos y + \cos z = 0 \qquad (11)$$

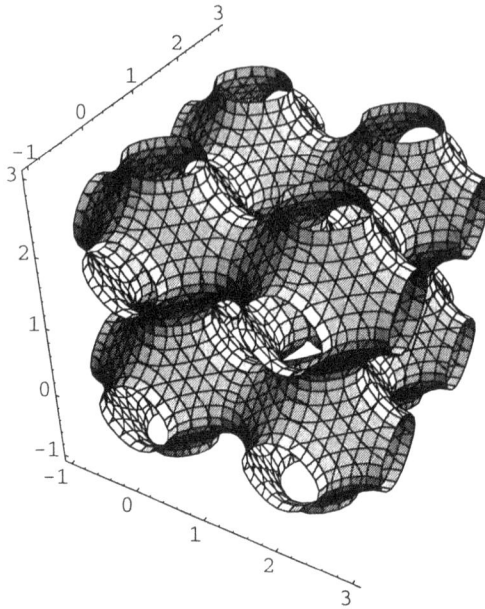

Fig 8. The 3D cos nodal surface very similar to the P-minimal surface.

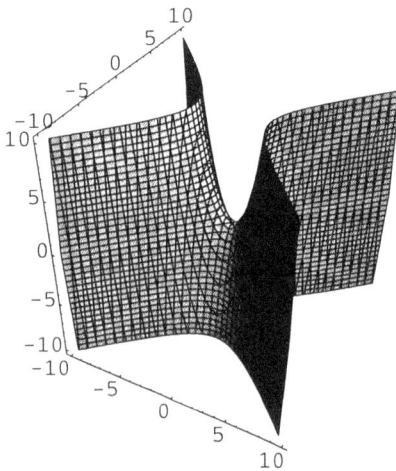

Fig 9a. Two planes meeting without intersections give the saddle surface.

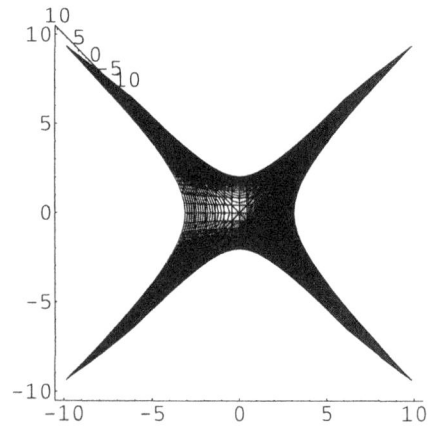

b. Different view of a.

The mathematics of the 3- periodic minimal surfaces is immensely more complicated than that of the nodal surfaces.

In figures 9 and 10 we show you what a saddle and a monkey saddle really are. The first saddle is the same as above but much more of it is shown. So we can see that it really consists of two planes that are going through each other without intersecting. This is shown in 9a and b.

The monkey saddle as shown in fig 10a and b is also as above, but a lot more. This time there are three planes that go through each other without intersecting, and the three fold symmetry is clear.

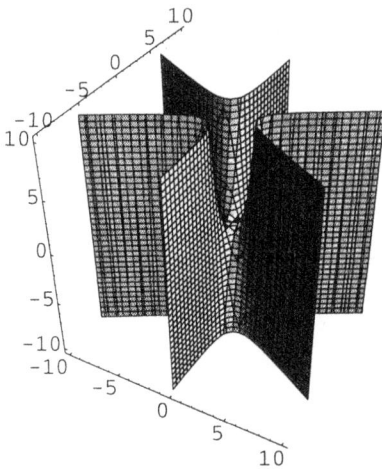

Fig 10a. Three planes meeting without intersections give the monkey saddle.

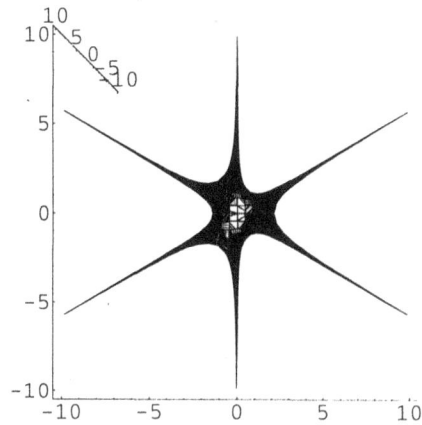

b. Different view of a.

We think that a great deal of the mathematics consists of how planes behave in its space. Surely, as you will see, the plane is a building unit not only in mathematics but also in chemical structures. Sometime planes intersect, sometime not. Sometime they form polyhedra, and sometime surfaces, If embedded we use them to describe molecules or ions or clusters. If planes form periodic surfaces we use them to describe crystal structures.

Planes according to our mathematics build polyhedra and by modulating them we get bodies that form simple structures. By changing a constant in the function the structure shrinks, and the bodies finally react and form new structures or surfaces. The mathematics here offers a unique picture as all the bodies move simultaneously. As they did in our description of the martensite transition[12]. This we call interaction, and use to describe bonding.

Physically and mathematically a crystal contains an endless number of planes, built up by the atoms. But electrons don't move in intersecting planes, they move in continuos and non-intersecting surfaces, and we believe we have the mathematics of such surfaces. You will see for yourself reading the book.

We do not include the fivefold symmetry, although it surely would fit in chapter 8. The field is huge, and almost untouched. It is enough to say that one of us has seen a surface in a model of a quasi-crystal structure[8].

LITERATURE

The common jargon of describing structures as atomic arrangements, as interconnected polyhedra, or as rod packing, or as nets has been described by us and others [2,9,]. The use of intrinsic curvature to describe various properties in solid state was introduced by us [10,11,12,13].

For the description and use of nodal and equipotential surfaces we refer to von Schnering and Nesper[14,15].

A giant work is presented by Hildebrandt et al in their two volumes Minimal surfaces I and II [16]. This must be by far the best in the field.

The use of minimal surfaces in lipid science is excellently covered by K. Larsson[17].

The work on the exponential scale is so far limited to three publications[18,19,20].

CHAPTER 1

POLYHEDRA

This is a book about the *Exponential Scale*. We shall avoid advanced mathematics, partly to make it readable and useful, and also because we do not worry much about it. Instead we will show you a *mathematical nature,* enormously rich and beautiful. But always simple. The tool we use is called experimental mathematics. We start at once;

$$x = 0 \qquad\qquad (1)$$

and

$$y = 0 \qquad\qquad (2)$$

are two planes that intersect in space. If written

$$10^x + 10^y = \text{const.} \qquad\qquad (3)$$

one plane continues into another, as shown in fig 1 below. This is what it is all about. We have obtained an analytic, continuous function by a lift to the power of 10. The base could be any positive number (except 0 or 1), of course. We have found the *rule of addition,* which says that different functions add to become one in a continuos, predictable manner. This rule seems to be unique for the exponential scale and makes this mathematics so useful. That is a fundament of this study. We shall see many remarkable examples of this rule ahead.

The constant (const.) in the equation is the value for which the function is displayed, the *isosurface constant*. Each function consists of a sea of function values distributed all over space. The symmetry of the distribution is determined by the different exponential parts in the function, and the isosurface constant determines which value is to be regarded in plotting the surface. Each function thus consists of a virtually infinite amount of surfaces, of which we only can display one at a time.

Now we add a third plane (eq. (4)), which is shown in fig 2.

$$10^x + 10^y + 10^{-x} = \text{const.} \qquad\qquad (4)$$

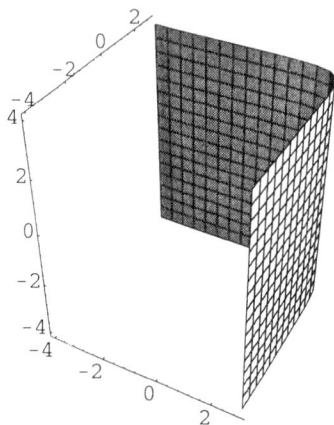

Fig 1. How two planes meet in space according to eq. $10^x + 10^y = $ const.

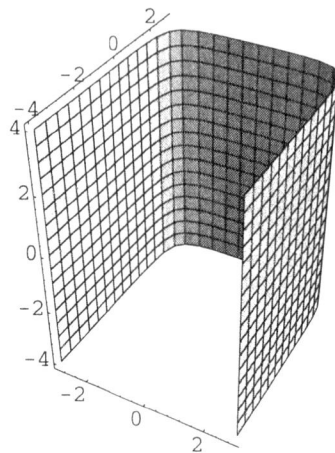

Fig 2. Three planes after eq. $10^x + 10^y + 10^{-x} = $ const.

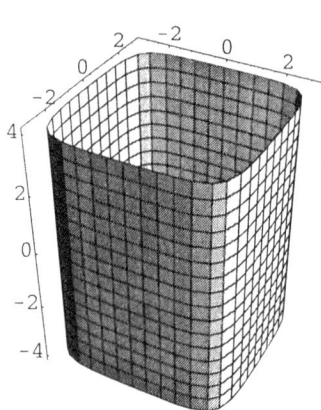

Fig 3. Four planes after eq. $10^x + 10^y + 10^{-x} + 10^{-y} = $ const.

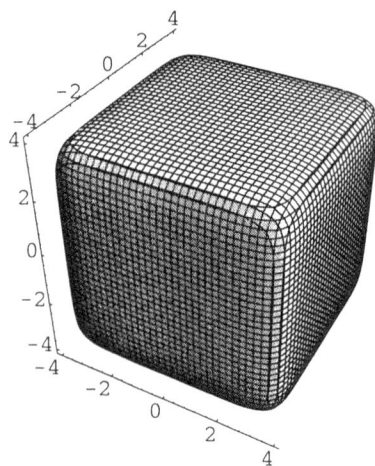

Fig 4. Six planes on the exponential scale forming a cube.

And another plane (fig 3). Now we have a column or a square tube. We put the lids on:

$$10^x + 10^y + 10^z + 10^{-x} + 10^{-y} + 10^{-z} = \text{const} \qquad (5)$$

and we got a cube (fig 4). We have changed the constant to 10000 from earlier 1000, to make the cube bigger, and sharper, and we have also increased the resolution.

Can we make the rest of the Platonic solids? Yes, of course we can. What we have used is not really planes, but face vectors. Such ones we easily find in for example Danas' Textbook of Mineralogy, the mineralogists call them the indices of faces. We can then calculate all sorts of crystal shapes. You can invent your own polyhedra and calculate them and you can also orient them any way you like which is useful. We will show you with the tetrahedron.

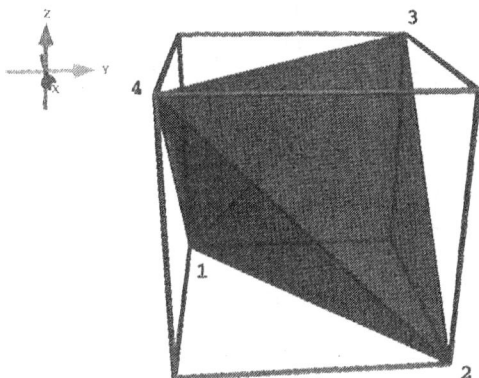

Fig 5a. A suitable orientation of a tetrahedron.

Place a tetrahedron in a coordinate system, for example as shown in fig 5a. The origin is in the centre of the outlined cube and the vertices (v) for the tetrahedron are:

$$v_1 = (-1 \ -1 \ -1) \qquad (6)$$
$$v_2 = (1 \ 1 \ -1) \qquad (7)$$
$$v_3 = (-1 \ 1 \ 1) \qquad (8)$$
$$v_4 = (1 \ -1 \ 1) \qquad (9)$$

In order to determine the face vectors we need the normal vectors to the four faces. Each face is defined by three vertices;

face 1: v_1, v_2, v_3
face 2: v_1, v_3, v_4
face 3: v_1, v_4, v_2
face 4: v_2, v_4, v_3

We need five of the six tetrahedrons edges (e) to determine the normals, and the vectors for these are:

$$e_{12} = v_2 - v_1 = (2\ 2\ 0) \qquad (10)$$
$$e_{13} = v_3 - v_1 = (0\ 2\ 2) \qquad (11)$$
$$e_{14} = v_4 - v_1 = (2\ 0\ 2) \qquad (12)$$
$$e_{23} = v_3 - v_2 = (-2\ 0\ 2) \qquad (13)$$
$$e_{24} = v_4 - v_2 = (0\ -2\ 2) \qquad (14)$$

The normals (n) are then the vector product of two edge vectors for the face (note that they are multiplied counter-clockwise in order to get the correct sign, or direction, of the normals);

$$n_1 = e_{13} \times e_{12} = (-4\ 4\ -4) \qquad (15)$$
$$n_2 = e_{14} \times e_{13} = (-4\ -4\ 4) \qquad (16)$$
$$n_3 = e_{12} \times e_{14} = (4\ -4\ -4) \qquad (17)$$
$$n_4 = e_{23} \times e_{24} = (4\ 4\ 4) \qquad (18)$$

For any face (even square, pentagonal or higher polygons) it is sufficient with just three vertices for the normal vector calculation, as three points define a plane.

And finally, to determine the face vectors we need a scale factor (s) for the distance of the face to the origin. This is calculated by scalar multiplication of the normal vector with an arbitrary vector to the face. For the Platonic solids the scale factors are naturally the same for all faces, but in this example we still calculate them all, and as our arbitrary vectors, we just choose one of the faces vertex vectors;

$$s_1 = n_1 \cdot v_1 = 4 \qquad (19)$$
$$s_2 = n_2 \cdot v_1 = 4 \qquad (20)$$
$$s_3 = n_3 \cdot v_1 = 4 \qquad (21)$$
$$s_4 = n_4 \cdot v_2 = 4 \qquad (22)$$

Now, to calculate the face vectors (u), we divide the normal vectors with the corresponding scale factor, and get;

$$u_1 = (-1 \ 1 \ -1) \tag{23}$$
$$u_2 = (-1 \ -1 \ 1) \tag{24}$$
$$u_3 = (1 \ -1 \ -1) \tag{25}$$
$$u_4 = (1 \ 1 \ 1) \tag{26}$$

The exponential scale equation for the tetrahedron is thus:

$$10^{tetr} = 10^{u_1 \cdot (x \ y \ z)} + 10^{u_2 \cdot (x \ y \ z)} + 10^{u_3 \cdot (x \ y \ z)} + 10^{u_4 \cdot (x \ y \ z)} =$$
$$10^{-x+y-z} + 10^{-x-y+z} + 10^{x-y-z} + 10^{x+y+z} = C \tag{27}$$

This method of face vector derivation is general for all polyhedra, and you can also scale and orient them as you like. For the tetrahedron, octahedron, the icosahedron, and dodecahedron we derive accordingly the following equations, with a short notation obvious from below:

$$10^{tetr} = 20000 \text{ (fig 5b);} \tag{28}$$

vectors: $(1 1 1), (1 \bar{1} \bar{1}), (\bar{1} \bar{1} 1), (\bar{1} 1 \bar{1})$

$$10^{oct} = 2 \cdot 10^6 \text{ (fig 6);} \tag{29}$$

vectors: $(\pm 1, \pm 1, \pm 1)$

$$10^{ico} = 10^{19} \text{ (fig 7);} \tag{30}$$

vectors: $(\pm\tau, \pm\tau, \pm\tau), (\pm\tau^2, 0, \pm 1), (\pm 1, \pm\tau^2, 0), (0, \pm 1, \pm\tau^2)$

$$10^{dod} = 10^{12} \text{ (fig 8);} \tag{31}$$

vectors, $(\pm\tau, \pm 1, 0), (\pm 1, 0, \pm\tau), (0, \pm\tau, \pm 1)$

$\frac{\sqrt{5}-1}{2} \approx 1.618$ is the golden section, τ, and 2.618 is $\tau+1$, or τ^2.

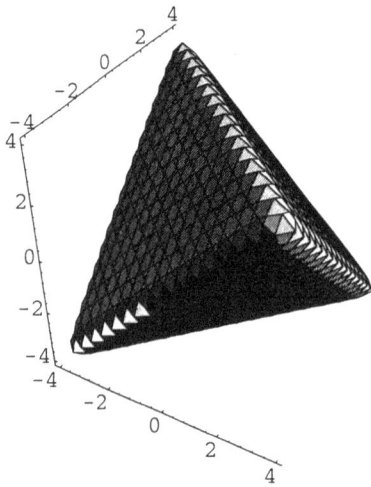

Fig 5b. Tetrahedron on the exponential scale with eq.
$10^{x+y+z}+10^{x-y-z}+10^{-x-y+z}+10^{-x+y-z}=20000$.

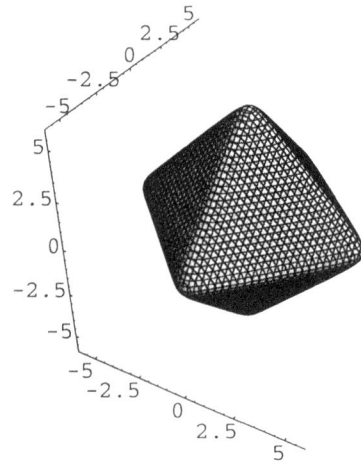

Fig 6. Octahedron.

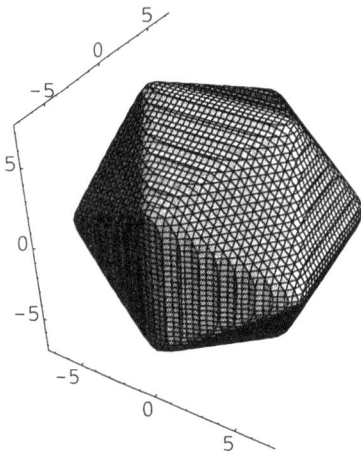

Fig 7. Icosahedron.

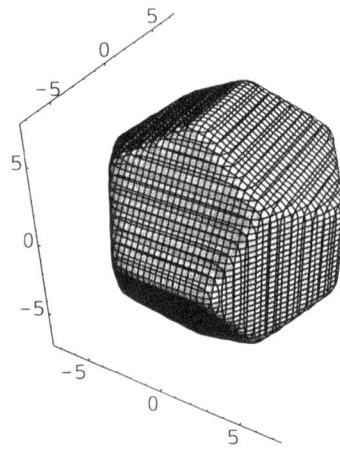

Fig 8. Dodecahedron.

It is clear we can do any polyhedra now. But let us concentrate on some which are important in chemistry. One is the cube octahedron, the fundamental unit of cubic close packing. Another is its hexagonal variant of equal importance in hexagonal close packing. A third is the rhombic dodecahedron, the fundamental unit of a bodycentered cubic structure. A fourth is the truncated tetrahedron. All four are shown below. By adding the equations of polyhedra by the law of *addition*, new truncations are obtained and/or other more complex polyhedra.

Figures 9, 10, 11 and 12 are resp. cube octahedron, its hexagonal correspondence, the rhombic dodecahedron, and the truncated tetrahedron. The equations are in order:

Cubeoctahedron:

$$10^{cube} + 10^{0.5oct} = 10^{15} \qquad\qquad (32)$$

Hexagonal cube octahedron:

$$10^{hex\_cuboct} = 10^{19}; \qquad\qquad (33)$$

vectors: $(\pm 2x, -\frac{2}{3}\sqrt{3}\, y, \pm\sqrt{\frac{2}{3}}\, z)$, $(\pm 2x, +\frac{2}{3}\sqrt{3}\, y, \pm 2\sqrt{\frac{2}{3}}\, z)$,

$(0, \frac{4}{3}\sqrt{3}\, y, \pm\sqrt{\frac{2}{3}}\, z)$, $(0, -\frac{4}{3}\sqrt{3}\, y \pm 2\sqrt{\frac{2}{3}}\, z)$, $(0,0,\pm 3\sqrt{\frac{2}{3}}\, z)$

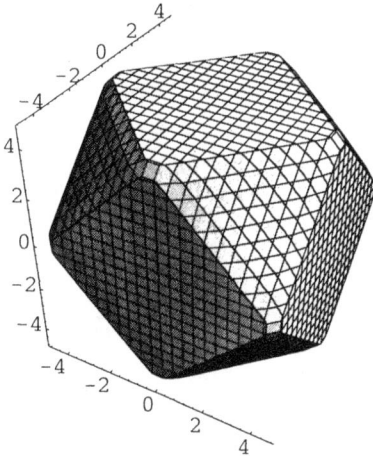

Fig 9. Cubeoctahedron.

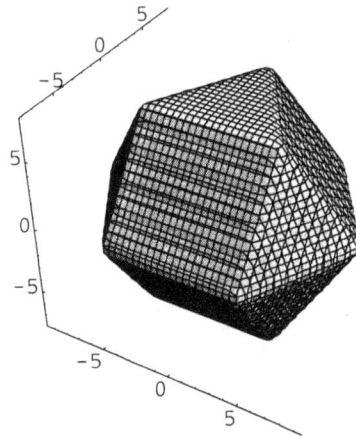

Fig 10. Hexagonal cube octahedron.

Rhombic dodecahedron:

$$10^{rho\_dod} = 10^{16} \tag{34}$$

vectors: $(\pm1, \pm1, 0), (\pm1, 0, \pm1), (\pm1, 0 \pm1)$

Truncated tetrahedron:

$$10^{tetr} + 10^{-2tetr} = 10^{8} \tag{35}$$

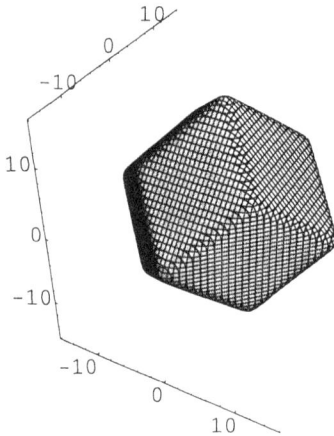

Fig 11. Rhombic dodecahedron.

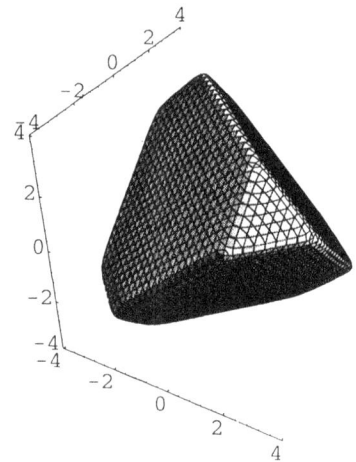

Fig 12. Truncated tetrahedron.

Can we do minerals? We have already done it, as you have seen in our polyhedra. A difficult one is the Pyritohedron, a form of pyrite. We know it is a distorted variant of the dodecahedron and look it up in *Dana* and find its face vectors and formulate the equation:

$$10^{pyrite} = 10^{12} \tag{36}$$

vectors: $(\pm2, \pm1, 0), (\pm1, 0, \pm2), (0, \pm2, \pm1)$

This equation is analogous to the dodecahedron as the pyritohedron, according to *Dana*, also has twelve faces. We see that the difference in symbol form is tau and 2 - Nature simply wants to make this crystal form of pyrite as close to the dodecahedron as possible but in a commensurate way.

Accordingly we get the shape in fig 13. Fig 14 is the dodecahedron.

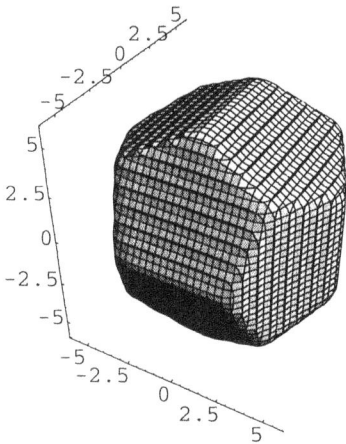

Fig 13. Pyritohedron. Vectors ($\pm 2, \pm 1, 0$), ($\pm 1, 0, \pm 2$), ($\pm 0, \pm 2, \pm 1$).

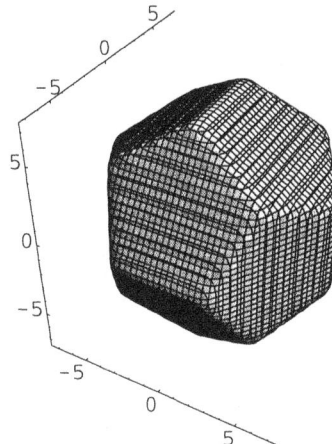

Fig 14. Dodecahedron. Vectors ($\pm\tau, \pm 1, 0$), ($\pm 1, 0, \pm \tau$), ($\pm 0, \pm\tau, \pm 1$).

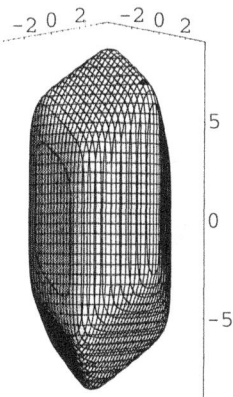

Fig 15a. Calcite crystal shape as calculated with truncations, guided by **Dana**.

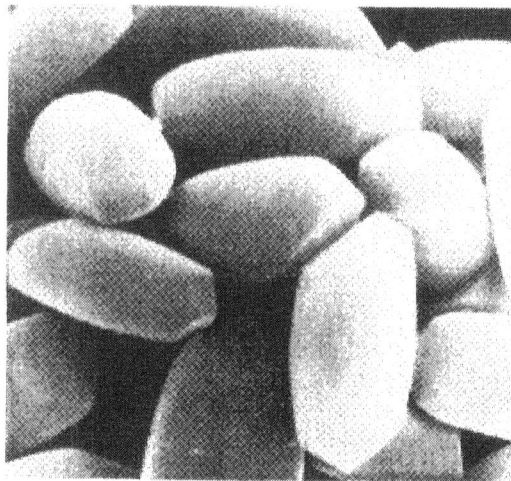

Fig 15b. The gravity sensor consists of small crystals of calcite.

Another strange mineral is a human one, the gravity sensor. This is small crystals of calcite and is found in our ears to help with the balance[1]. With *Dana* and various truncations on calcite shapes we have calculated fig 15a. As comparison a picture of the real crystals is shown in fig 15b.

Curvature

The larger the isosurface constant for the polyhedra is, the sharper they get. But how does the size of the polyhedra actually change the curvature? Lets take a look at the cube. For calculating the Gaussian and mean curvature we use the Mathematica script supplied in appendix 1.

With the base e, the equation for the cube is

$$e^x + e^y + e^z + e^{-x} + e^{-y} + e^{-z} = C. \tag{37}$$

We simplify the expressions of the curvatures by looking at three special cases, a vertex, an edge and the middle of a face.

At a vertex the coordinates are one of the eight permutations of x=y=z, and the Gaussian curvature is such a point is

$$K = \frac{2 - \dfrac{1}{e^{4x}} - e^{4x}}{(-e^{-x} + e^x)^2 (6 - \dfrac{3}{e^{2x}} - 3e^{2x})} \tag{38}$$

and the mean curvature

$$H = \frac{-\sqrt{3}(\dfrac{1}{e^{3x}} - \dfrac{1}{e^x} - e^x + e^{3x})}{(-e^{-x} + e^x)(6 - \dfrac{3}{e^{2x}} - 3e^{2x})} \tag{39}$$

The free variable, x, is the size of the cube. From this we see that both the curvatures decrease with increased size, to become constant for larger cubes. The Gaussian curvature converges to $K = \dfrac{1}{3}$ and the mean to $H = \dfrac{1}{\sqrt{3}}$, which is seen in diagrams 1a and b.

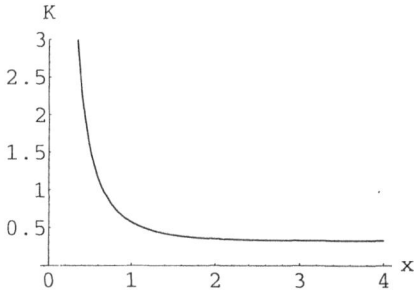

Diagram 1a. Gaussian curvature of a corner of the exponential scale cube.

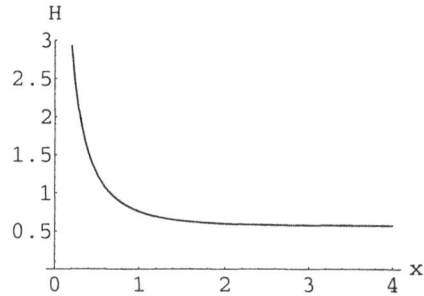

Diagram 1b. Mean curvature of a corner.

At an edge, x=y, while z=0. The curvatures at this point are

$$K = \frac{2(\dfrac{-1}{e^{3x}} + \dfrac{1}{e^x} + e^x - e^{3x})}{(-e^{-x} + e^x)^2(4 - \dfrac{2}{e^{2x}} - 2e^{2x})} \tag{40}$$

$$H = \frac{-(-4 + \dfrac{1}{e^{3x}} + \dfrac{2}{e^{2x}} - \dfrac{1}{e^x} - e^x - 2e^{2x} + e^{3x})}{\sqrt{2}(-e^{-x} + e^x)(4 - \dfrac{2}{e^{2x}} - 2e^{2x})} \tag{41}$$

These plots are shown in diagram 2a and b.

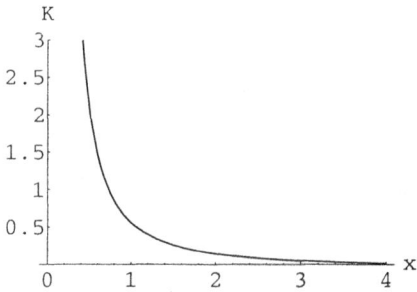

Diagram 2a. Gaussian curvature of an edge.

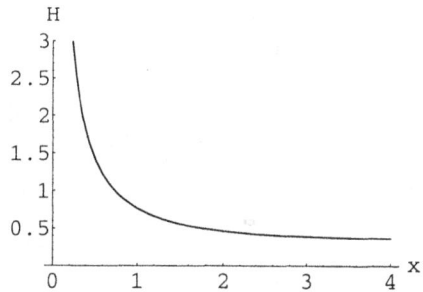

Diagram 2b. Mean curvature of an edge.

As expected the Gaussian curvature converges to zero, since the edge is similar to a cylinder and thus has parabolic geometry. The mean curvature converges to $\frac{1}{2\sqrt{2}}$.

At the middle of a face the two curvatures converge to zero because the face turns more and more planar. The curvatures are

$$K = \frac{8 - \dfrac{4}{e^{2x}} - 4e^{2x}}{(-e^{-x} + e^x)^2 (2 - e^{-2x} - e^{2x})} \qquad (42)$$

$$H = \frac{4 - \dfrac{2}{e^{2x}} - 2e^{2x}}{(-e^{-x} + e^x)(2 - e^{-2x} - e^{2x})} \qquad (43)$$

and their plots are shown in diagram 3a and b.

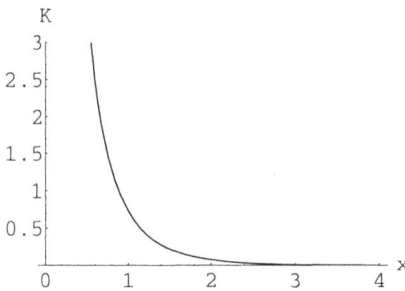

Diagram 3a. Gaussian curvature of a face in the exponential scale cube.

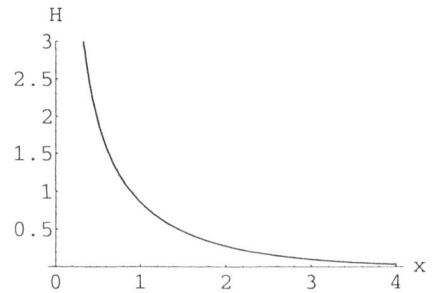

Diagram 3b. Mean curvature of a face.

Thus, all polyhedra in the exponential scale converge their Gaussian and mean curvatures when they grow with the isosurface constant. At faces, both of them are zero, and at edges the Gaussian curvature is zero.

The smaller the polyhedra, the lower the constant and the more each vertex gets affected by the others, which results in the polyhedra turning spherical and the curvatures increase.

THE NATURAL FUNCTION

In the exponential scale we can chose any base, as earlier mentioned - using e makes it easy to demonstrate the link to traditional functions.

e^x is a fundamental function in mathematics. It is the sum of the hyperbolic functions:

$$e^x = \cosh x + \sinh x \tag{44}$$

where

$$\cosh x = \frac{1}{2}(e^x + e^{-x}) \tag{45}$$

and

$$\sinh x = \frac{1}{2}(e^x - e^{-x}) \tag{46}$$

What is the geometry? We shall see it is beautiful in three dimensions - we have already seen that *cosh* is elliptic and gave us all the polyhedra and the crystals. Let us do the power expansions:

$$\cosh x + \cosh y + \cosh z = 3 + \frac{1}{2!}(x^2 + y^2 + z^2) + $$
$$\frac{1}{4!}(x^4 + y^4 + z^4) \tag{47}$$

It is clearly elliptic geometry, and in fig 16 we see the function for the terms used above.

Clearly it will be more cubic with more terms.
What happens if we study the other, *sinh* ? The expansion is:

$$\sinh x + \sinh y + \sinh z = $$
$$x + y + z + \frac{1}{3!}(x^3 + y^3 + z^3) + \frac{1}{5!}(x^5 + y^5 + z^5) \tag{48}$$

and in fig 17 we have calculated the three first terms. The result is a monkey saddle - the arch type of hyperbolic geometry!

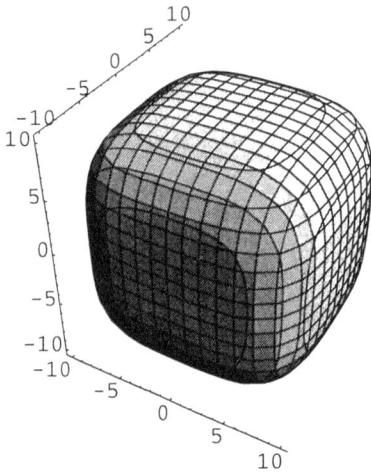

Fig 16. Power expansion of cosh(cubic) in 3D.

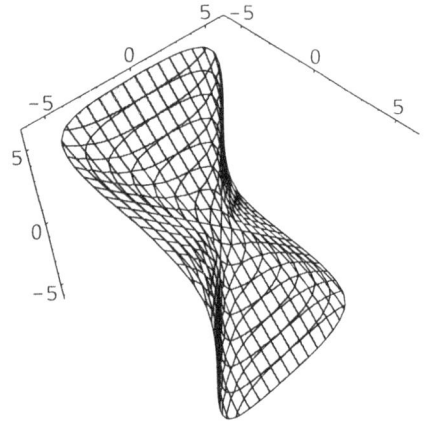

Fig 17. Power expansion of sinh(cubic) in 3D.

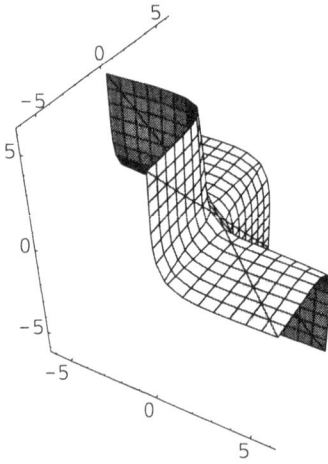

Fig 18. Power expansion of sinh(oct) in 3D.

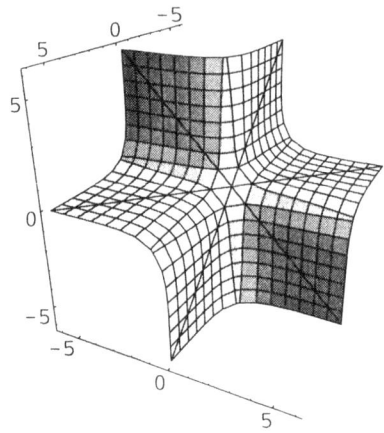

Fig 19. Different view from 18.

Next time we use octahedral planes in an expansion:

$$\sinh(oct) = 2x + 2y + 2z + \frac{1}{3!}((-x+y+z)^3 +$$

$$(x+y-z)^3 + (x-y+z)^3 + (x+y+z)^3) +$$

$$\frac{1}{5!}((x+y+z)^5 + (x+y-z)^5 + (y+z-x)^5 + (x-y+z)^5)$$

$$(49)$$

And a very beautiful monkey saddle is the result and shown in figs 18 and 19.

Now we shall look on a fifty - fifty sum of the two which of course is

$$e^x + e^y + e^z \qquad\qquad (50)$$

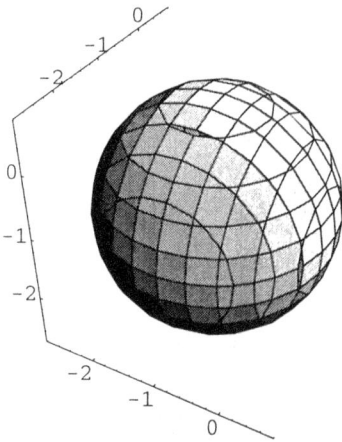

Fig 20. Power expansion of the natural function. Two terms.

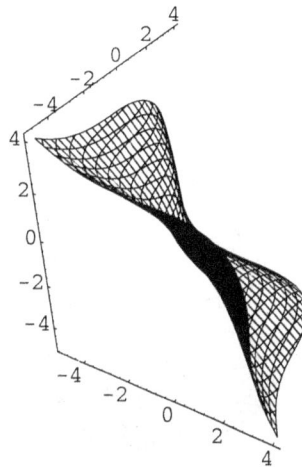

Fig 21. Three terms.

We have stepwise plotted this function up to nine terms in the expansion:

$$x + y + z + \frac{1}{2!}(x^2 + y^2 + z^2) = 0, \qquad (51)$$

shown in fig 20.

$$3 + x + y + z + \frac{1}{2!} (x^2 + y^2 + z^2) + \frac{1}{3!} (x^3 + y^3 + z^3) = 0, \tag{52}$$

shown in fig 21.

$$x + y + z + \frac{1}{2!} (x^2 + y^2 + z^2) + \frac{1}{3!} (x^3 + y^3 + z^3) +$$
$$\frac{1}{4!} (x^4 + y^4 + z^4) = 0, \tag{53}$$

shown in fig 22.

$$x + y + z + \frac{1}{2!} (x^2 + y^2 + z^2) + \frac{1}{3!} (x^3 + y^3 + z^3) +$$
$$\frac{1}{4!} (x^4 + y^4 + z^4) + \frac{1}{5!} (x^5 + y^5 + z^5) + 3 = 0, \tag{54}$$

shown in fig 23.

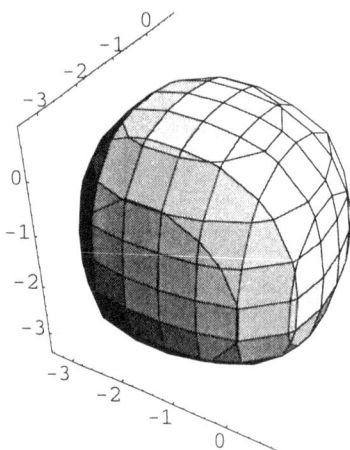

Fig 22. Four terms.   **Fig 23.** Five terms.

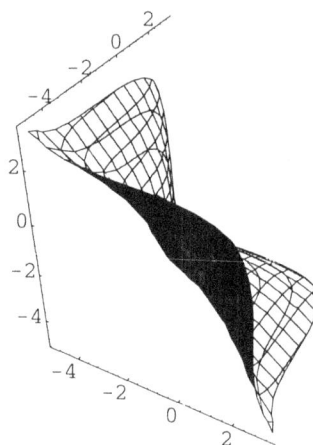

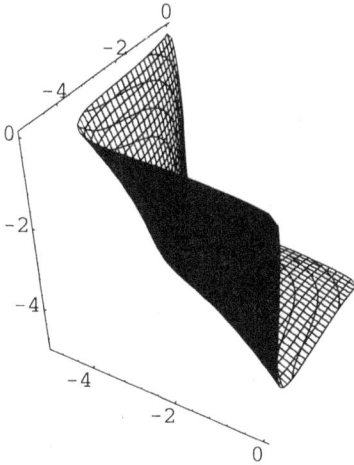

Fig 24. Nine terms.   **Fig 25.** Ten terms.

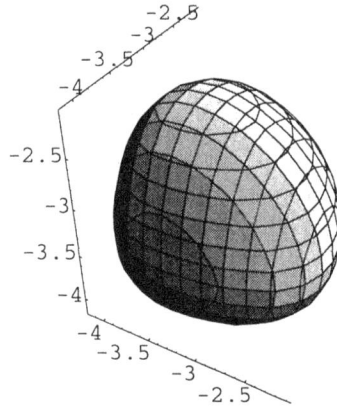

Figures 24 and 25 corresponds to terms 9 and 10 in the expansion.

What we see is a solid rock of mathematics in one single function; the euclidean plane for the first term and an oscillation between elliptic and hyperbolic geometry for the next terms, until the end (a corner of a cube!). In fig 24 it is clear that the function also for these terms is becoming a cube corner.

Just one example of a non fifty - fifty sum of *sinh* and *cosh*.
We write the equation

$$\cosh x + \cosh y - \sinh z = \text{const} \tag{55}$$

and do the expansion, using the three first terms:

$$2 + \frac{1}{2!}x^2 + \frac{1}{4!}x^4 + \frac{1}{2!}y^2 + \frac{1}{4!}y^4 - z - \frac{1}{3!}z^3 - \frac{1}{5!}z^5 = 50 \tag{56}$$

which is shown in fig 26.

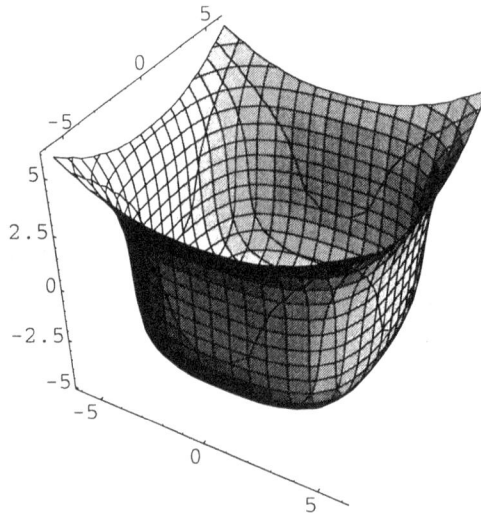

Fig 26. Subtraction opens a polyhedron via a catenoid.
Eq. coshx+coshy-sinhz = const.

We got a cube with one side opened up to a catenoid, and that is really the start of the next chapter.

Conclusion

The expansion of $e^x+e^y+e^z$ is an oscillation between elliptic and hyperbolic geometry.

If we plot $e^x+e^y+e^z$ we find a cube corner. We now realise that <u>cosh</u> is joining polyhedral corners via edges to elliptic geometry, <u>sinh</u> has the opposite orientation and corners meet to hyperbolic geometry. By mixing these two and using the law of addition we get continuous functions, polyhedra or monkey saddles or saddles, in the chapters to come. But with the jargon we use in the trade - building structures - it is easier to use the concepts of planes or face vectors.

CHAPTER 2

SADDLES

Subtraction in a polyhedral equation gives hyperbolic geometry.
The cube is changed to a monkey saddle:

$$10^x + 10^y + 10^z - 10^{-x} - 10^{-y} - 10^{-z} = 0 \qquad (1)$$

Notice that the constant now is zero. The monkey saddle, shown
in various orientations and magnifications in figs 1, 2, 3, and 4
contain lines, very close to straight, or two fold axes.

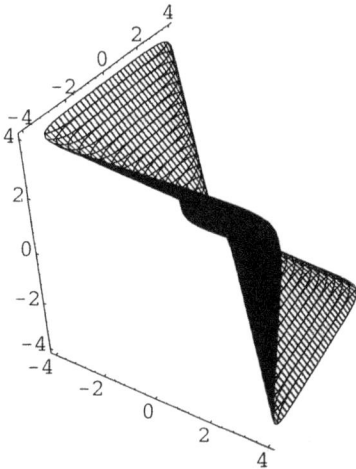

Fig 1. Changing signs in the cube equation
$10^x + 10^y + 10^z - 10^{-x} - 10^{-y} - 10^{-z} = 0$ make the
planes meet in a point and form a monkey
saddle.

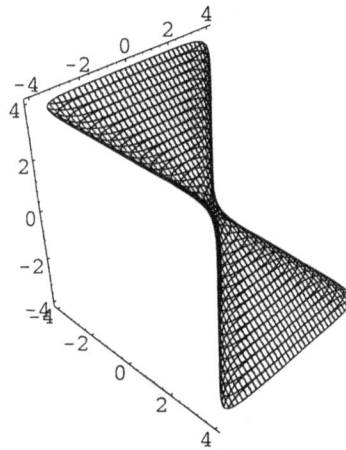

Fig 2. The monkey saddle and its two fold
axes.

The equation for the tetrahedron, modified to give hyperbolic
geometry is:

$$10^{-x+y+z} - 10^{x+y-z} + 10^{x-y+z} - 10^{-x-y-z} = 0 \qquad (2)$$

Naturally this becomes a saddle, shown in figs 5 and 6, in two
different magnifications.

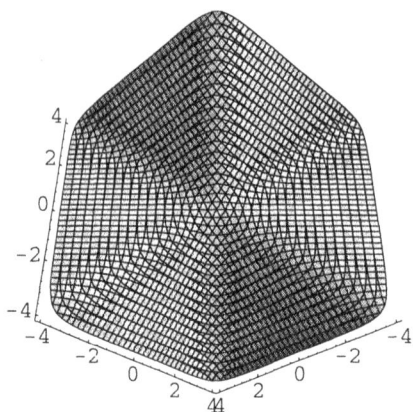

Fig 3. The monkey saddle and its three fold axes.

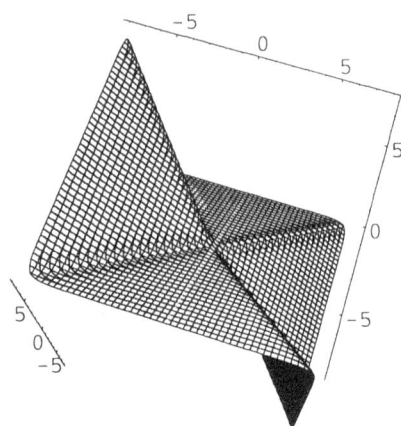

Fig 4. A bigger part of the monkey saddle clearly shows the planes.

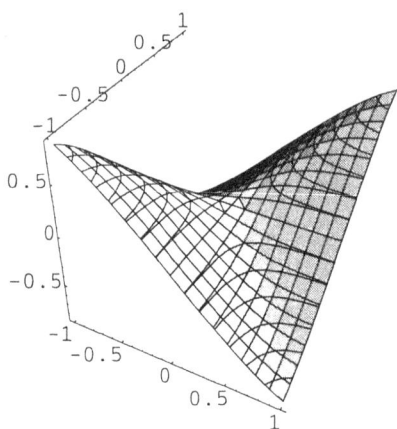

Fig 5. The tetrahedron gives a saddle:
$$10^{-x+y+z} - 10^{x+y-z} + 10^{x-y+z} - 10^{-x-y-z} = 0$$

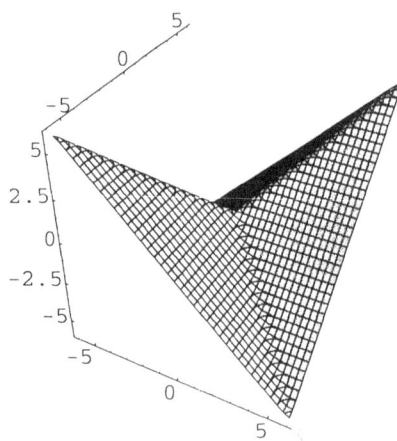

Fig 6. A bigger part shows the planes.

The octahedron gives a very simple saddle:

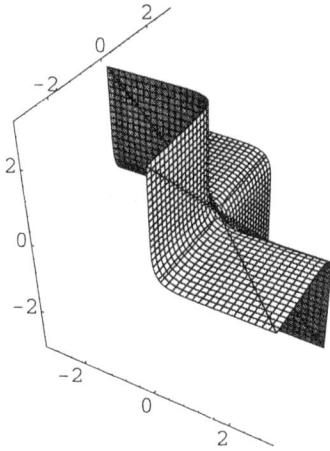

Fig 7. The octahedral monkey saddle.

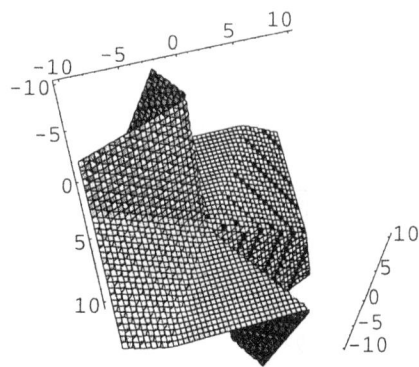

The icosahedron gives a monkey saddle shown in fig 8a and b.

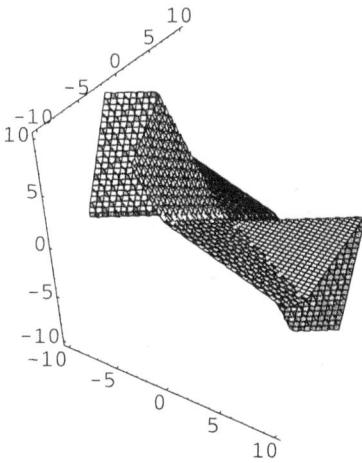

Fig 8a. The icosahedral monkey saddle.   **Fig 8b.** Different view of a.

The dodecahedron gives a magnificent monkey saddle composed of twelve planes, as shown in fig 9 and 10, the latter with high resolution.

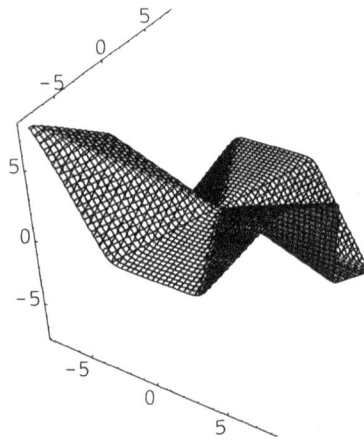

Fig 9. The dodecahedral monkey saddle.

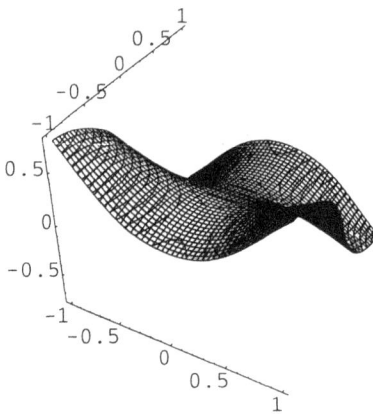

Fig 10. Magnification of saddle in Fig 9.

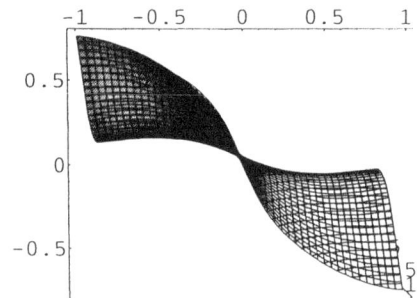

Fig 11. The two fold axis of the dodecahedral monkey saddle.

In the last figure, 11, we have rotated fig 10 to show one of the straight lines, or two folded axes.

We do one more, the rhombic dodecahedron, and here is the
equation:

$$10^{(x+y)} - 10^{-(x+y)} + 10^{(-x+y)} - 10^{-(-x+y)} +$$
$$10^{(y+z)} - 10^{-(y+z)} + 10^{(-y+z)} - 10^{-(-y+z)} + \qquad (3)$$
$$10^{(-x+z)} - 10^{-(-x+z)} + 10^{(x+z)} - 10^{-(x+z)} = 0$$

Below we see two different projections of this simple monkey
saddle, in figures 12a and b.

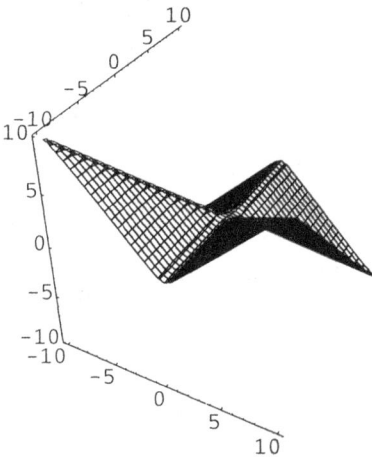

Fig 12a. The monkey saddle of the
rhombic dodecahedron.

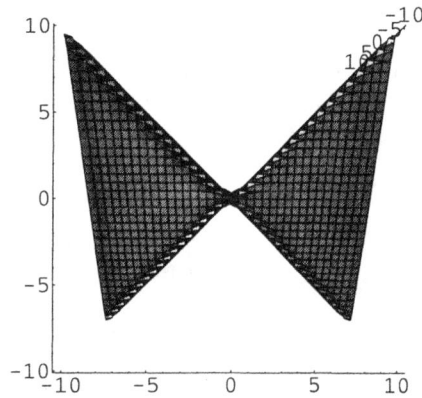

Fig 12b. Its two fold axis.

CATENOIDS

If we in the equation of a polyhedron change the sign of one
exponential term, we introduce saddles, and the corresponding
plane opens up to a catenoid.

$$10^x + 10^y - 10^z + 10^{-x} + 10^{-y} + 10^{-z} = \text{const} \qquad (4)$$

This is shown in fig 13.

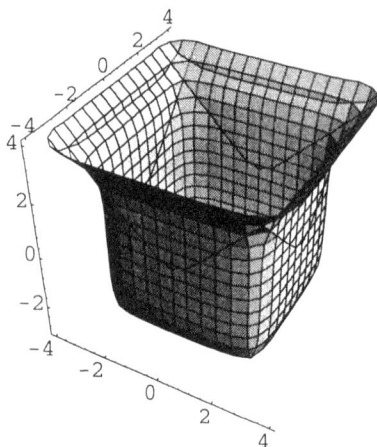

Fig 13. Changing one sign in the cube eq. $10^x + 10^y - 10^z + 10^{-x} + 10^{-y} + 10^{-z} = \text{const}$ gives catenoidic opening.

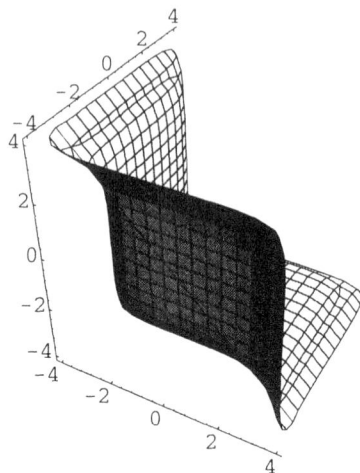

Fig 14. Two more faces opened.

We can open two more faces and have a splendid cube-catenoid centaur structure, as shown in fig 14. Of course this can be done on all polyhedra.

More beautiful structures are obtained by opening the corners of a polyhedron. This is done with the following trick: A polyhedron is truncated in each corner and one of the terms $(10^{(x+y+z)}$ below) is given a negative sign. We exemplify with the cube, and octahedral truncation:

$$
\begin{aligned}
&10^{(-x+y+z)} + 10^{(x+y-z)} + 10^{(x-y+z)} + \\
&10^{(-x-y-z)} - 10^{(x+y+z)} + 10^{-(x+y-z)} + \\
&10^{-(y+z-x)} + 10^{-(x-y+z)} + 10^{2.5x} + 10^{-2.5x} + \\
&10^{2.5y} + 10^{-2.5y} + 10^{2.5z} + 10^{-2.5z} = 1000
\end{aligned}
\tag{5}
$$

The corner with negative octahedral term $(10^{(x+y+z)})$ becomes a catenoid as is shown in fig 15.

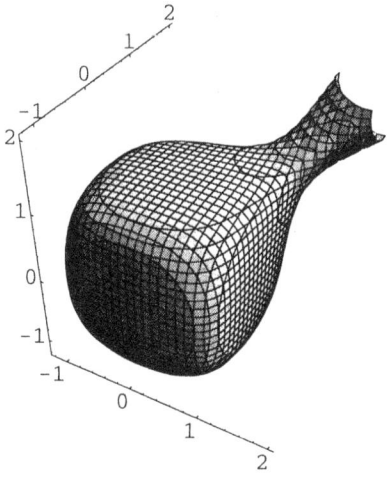

Fig 15. A cube corner is opened via making one of the terms negative in the octahedral truncation.

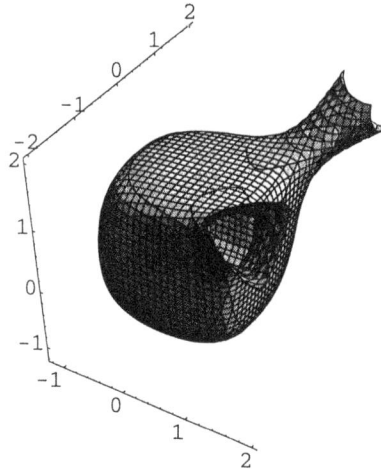

Fig 16. Two negative terms give two corner catenoids of the cube.

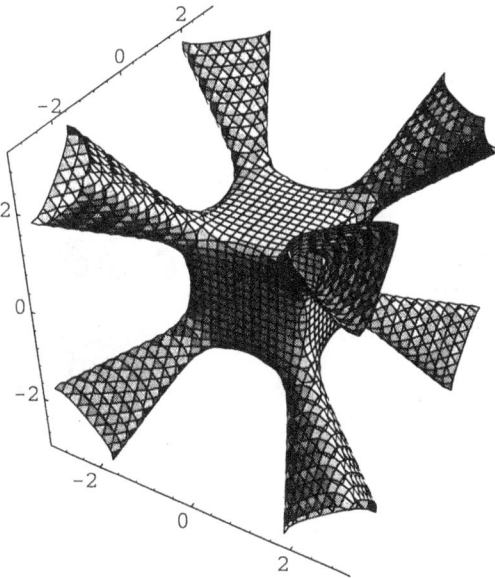

Fig 17. All eight corners opened.

We open one more in fig 16 and finally all eight in fig 17, with the simple equation

$$10^{2.5\text{cube}} - 10^{\text{oct}} = 1000 \tag{6}$$

Similarly, the tetrahedral truncation of the tetrahedron as in the equation below gives the very beautiful structure of fig 18.

$$10^{\text{tetr}} - 10^{-2\text{tetr}} = -60$$

or written out

$$10^{(-x+y+z)} + 10^{(x+y-z)} + 10^{(-x-y-z)} + 10^{(x-y+z)} -$$
$$10^{-(2(-x+y+z))} - 10^{-(2(x+y-z))} - 10^{-(2(x-y+z))} - \tag{7}$$
$$10^{-(2(-x-y-z))} + 60 = 0$$

By a cubic truncation of the octahedron and with the equation below we obtain fig 19.

$$10^{2\text{oct}} - 10^{3\text{cube}} = 1000 \tag{8}$$

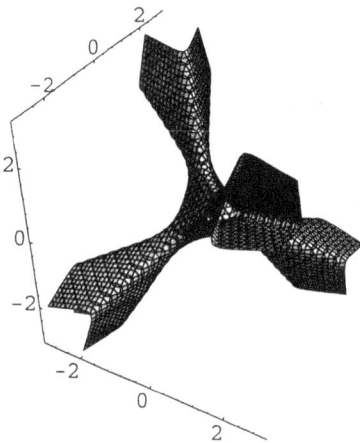

Fig 18. Negative tetrahedral truncation of the tetrahedron.

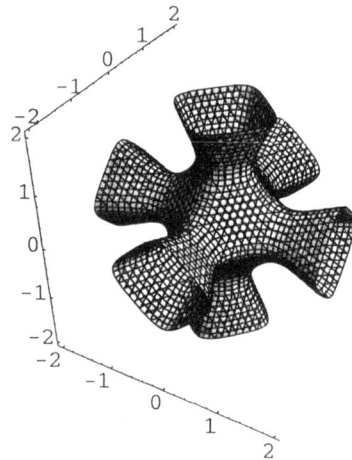

Fig 19. Negative cubic truncation of the octahedron.

Just making one corner catenoidic we get the 'hanging drop' shown in fig 20.

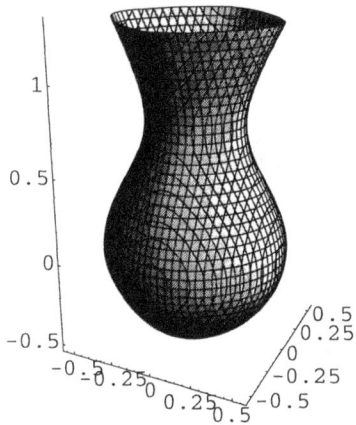

Fig 20. The hanging drop.

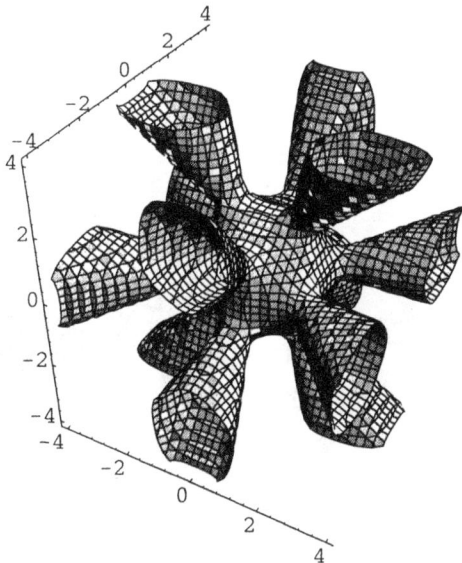

Fig 21. Negative dodecahedral truncation on the icosahedron.

Now we do the icosahedron with negative signs for the contributions of the dodecahedral truncations in the equation:

$$10^{ico} - 10^{1.4dod} = 100000 \tag{9}$$

and in fig 21 we see the formidable result of a formidable equation.

Finally we do the trinoid, which has no regular polyhedron as background. So we use the trigonal prism with its equation:

$$10^{(x+\frac{\sqrt{3}}{3}y)} + 10^{-\frac{2\sqrt{3}}{3}y} + 10^{(-x+\frac{\sqrt{3}}{3}y)} + 10^{0.7z} + 10^{-0.7z} = 1000 \tag{10}$$

and truncate it towards the hexagonal prism and open it up according to the equation:

$$10^{(x+\frac{\sqrt{3}}{3}y)} + 10^{-\frac{2\sqrt{3}}{3}y} + 10^{(-x+\frac{\sqrt{3}}{3}y)} + 10^{z} + 10^{-z} -$$
$$10^{-(x+\frac{\sqrt{3}}{3}y)} - 10^{\frac{2\sqrt{3}}{3}y} - 10^{-(-x+\frac{\sqrt{3}}{3}y)} = 10 \tag{11}$$

The result is shown in fig 23.

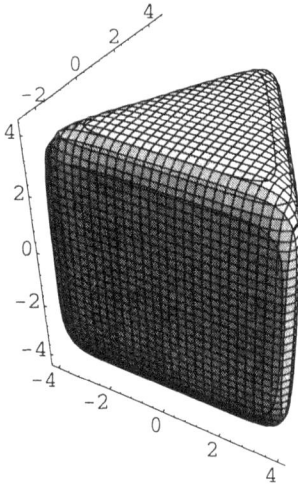

Fig 22. The trigonal prism.

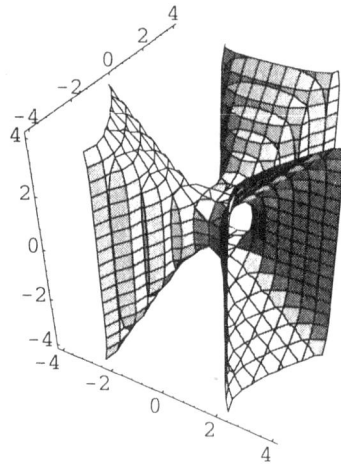

Fig 23. Negative hexagonal truncation of the prism gives the trinoid.

CHAPTER 3

PLANES, SPHERES AND CLUSTERS

So far we have only added planes together - now we add or subtract spheres to functions in our earlier work. This follows the rule of addition and adding a sphere to an open function means closing it. Working with opened polyhedra give two cases as we shall see.

The packing of spheres has always fascinated man - here we show how to join them with mathematics from *the exponential scale*.

We have found three different ways to join three spheres. In the first the trigonal prism is truncated along the edges towards the hexagonal prism and opened up according to the equation (1). In the second we have opened the triangular faces, and subtracted the sphere, giving equation (2). In the third case we have added a sphere to the trinoid. Below are the three equations and the corresponding figures 2, 3, and 4.

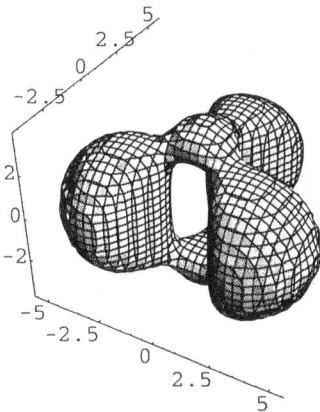

Fig 1. How to join three spheres after eq. (1).

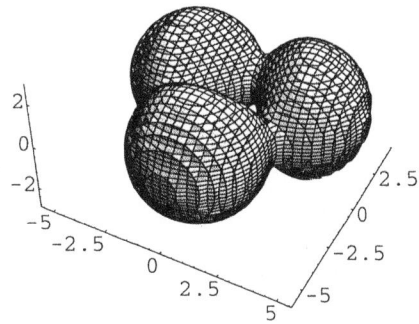

Fig 2. How to join three spheres after eq. (2).

$$10^{(x+\frac{\sqrt{3}}{3}y)} + 10^{-\frac{2\sqrt{3}}{3}y} + 10^{(-x+\frac{\sqrt{3}}{3}y)} - 10^{0.75z} -$$

$$10^{-0.75z} - 10^{-(x+\frac{\sqrt{3}}{3}y)} - 10^{\frac{2\sqrt{3}}{3}y} - 10^{-(-x+\frac{\sqrt{3}}{3}y)} + \qquad (1)$$

$$10^{0.2(x^2+y^2+z^2)} + 30 = 0$$

$$10^{(x+\frac{\sqrt{3}}{3}y)} + 10^{-\frac{2\sqrt{3}}{3}y} + 10^{(-x+\frac{\sqrt{3}}{3}y)} -$$

$$10^z - 10^{-z} - 10^{0.2(x^2+y^2+z^2)} = 10 \qquad (2)$$

$$10^{(x+\frac{\sqrt{3}}{3}y)} + 10^{-\frac{2\sqrt{3}}{3}y} + 10^{(-x+\frac{\sqrt{3}}{3}y)} +$$

$$10^z + 10^{-z} - 10^{-(x+\frac{\sqrt{3}}{3}y)} - 10^{\frac{2\sqrt{3}}{3}y} - \qquad (3)$$

$$10^{-(-x+\frac{\sqrt{3}}{3}y)} + 10^{0.3(x^2+y^2+z^2)} = 10$$

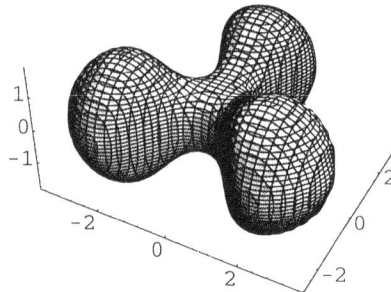

Fig 3. How to join three spheres.
A sphere is added to the trinoid.

In order to join four spheres we open two opposite faces of the cube and subtract a sphere:

$$10^{2.5x} + 10^{-2.5x} + 10^{2.5y} + 10^{-2.5y} -$$
$$10^{2.5z} - 10^{-2.5z} - 10^{(x^2+y^2+z^2)} = 10 \tag{4}$$

A kind of torus is seen in two different projections in fig 4 and 5

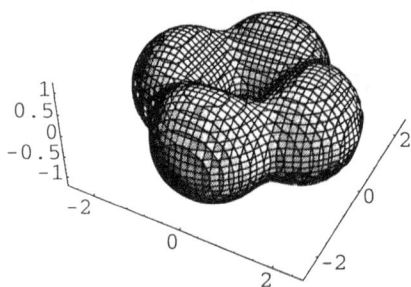

Fig 4. How to join four spheres. After eq. (4).

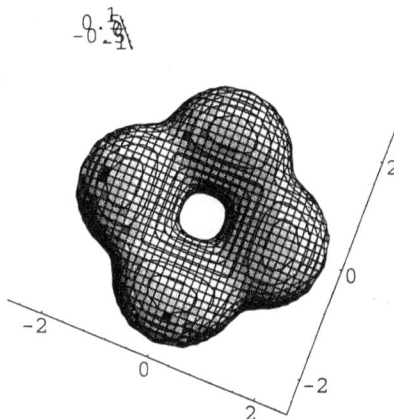

Fig 5. Different view of fig. 4.

Now we use the opened cube (fig 17, chapt 2) and add a large sphere:

$$10^{oct} - 10^{2.5cube} + 10^{0.33sphere} = -200 \tag{5}$$

In fig 6 we see that a cluster of *ccp* is formed. With smaller boundaries we see the inside of fig 6 in fig 7.

And in figs 8a and b we see split versions of fig 6. In chemistry we normally talk about the octahedral interstices in *ccp* but in this case it is the dual.

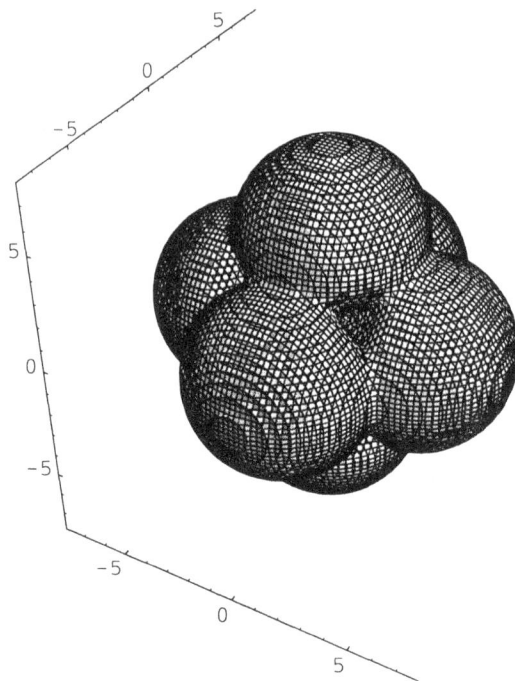

Fig 6. A sphere added to the opened cube of fig 17, chapter 2 .

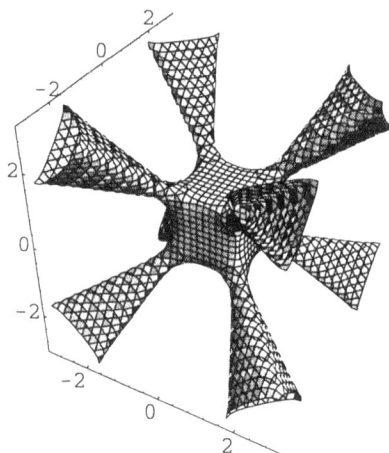

Fig 7. The inside of Fig 6.

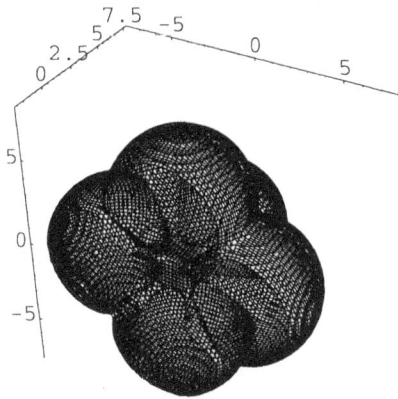

Fig 8a. A split of fig 6.

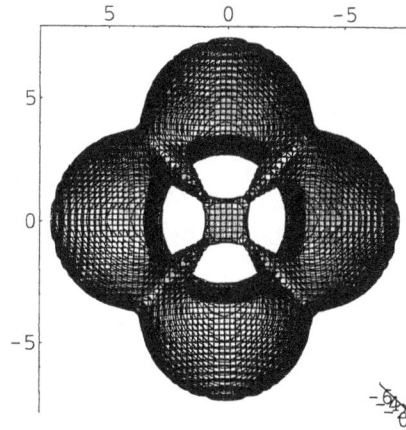

Fig 8b. Different orientation of a.

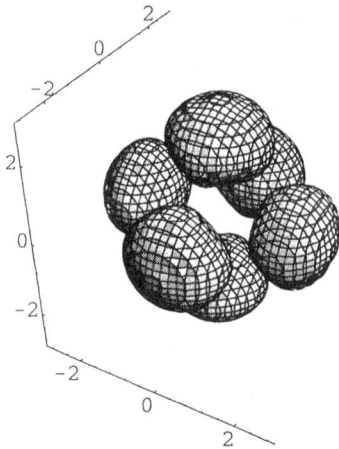

Fig 9a. The prehistory of fig 6. The
spheres have not yet met (eq. (6)).

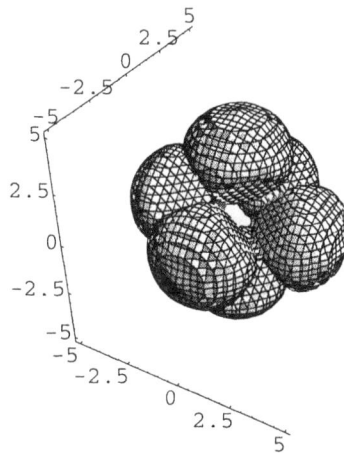

Fig 9b. Spheres just cluster to form fig 6
via catenoids according to eq. (7).

We can mimic another prehistory of fig 6. By varying constants
we can make free atoms approach, and interact to form the
cluster as in figs 9a and b. The equations are respectively:

$$10^{oct} - 10^{2cube} + 10^{0.7sphere} = -50 \qquad\qquad (6)$$

$$10^{oct} - 10^{2cube} + 10^{0.4sphere} = -50 \qquad\qquad (7)$$

What we have seen so far is catenoids combining in the folding up to form spheres.

We can also make them close up individually, as seen in fig 10, by varying the size of the sphere, and polyhedra. The equation is here:

$$10^{oct} - 10^{3cube} + 10^{2sphere} = 1000 \qquad\qquad (8)$$

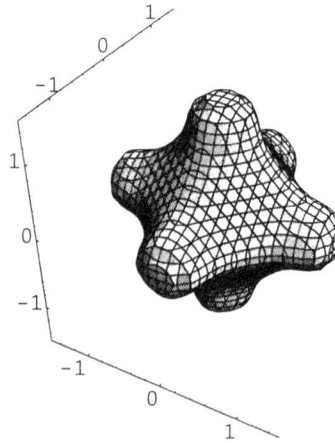

Fig 10. Another form of fig 6, from eq. (8).

We can also make analogous structures with above using the equation

$$10^{1.5cube} - 10^{oct} + 10^{0.33sphere} + 30 = 0 \qquad\qquad (9)$$

and the opened octahedron now forms a primitive packing of spheres as shown in fig 11 with its split version in fig 12a. Again we see that the interstices formed is the dual, which we show in fig 12b.

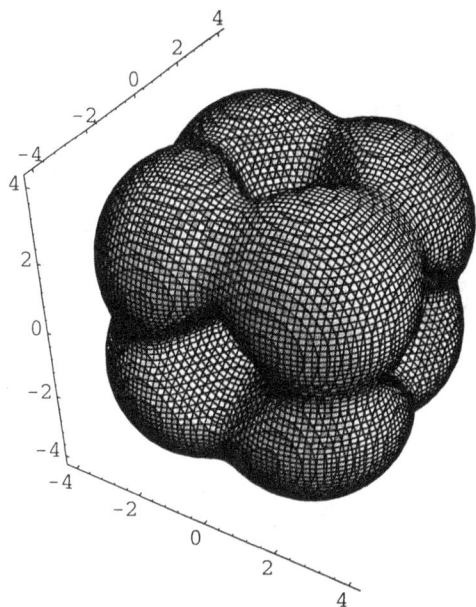

Fig 11. A sphere added to the opened octahedron of fig 19, chapter 2.

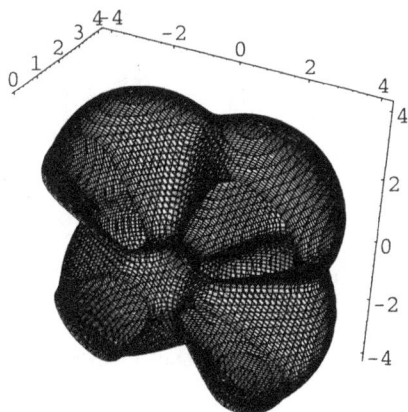

Fig 12a. A split of fig 11.

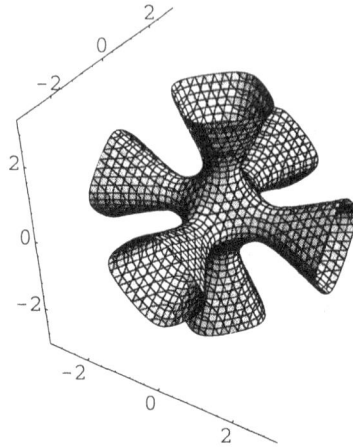

Fig 12b. The inside of fig. 11 (the opened octahedron).

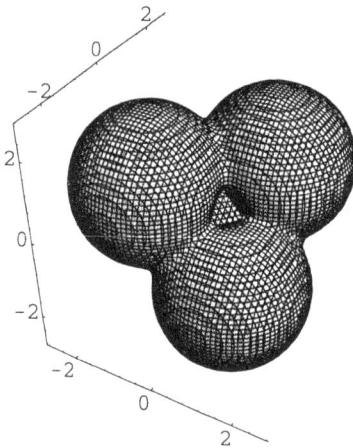

Fig 13. The tetrahedral cluster.

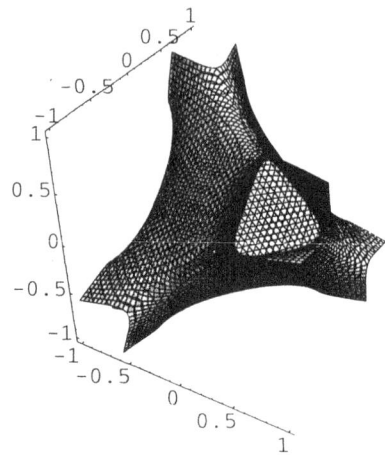

Fig 14. Its own dual, the inside of fig 13.

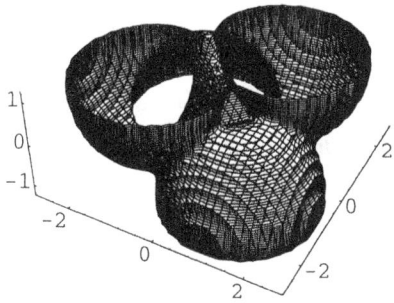

Fig 15. A split of fig 13.

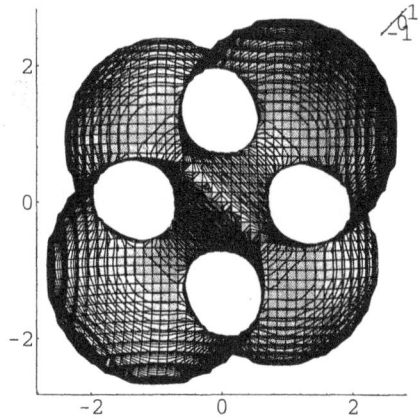

Fig 16. Another split of fig 13.

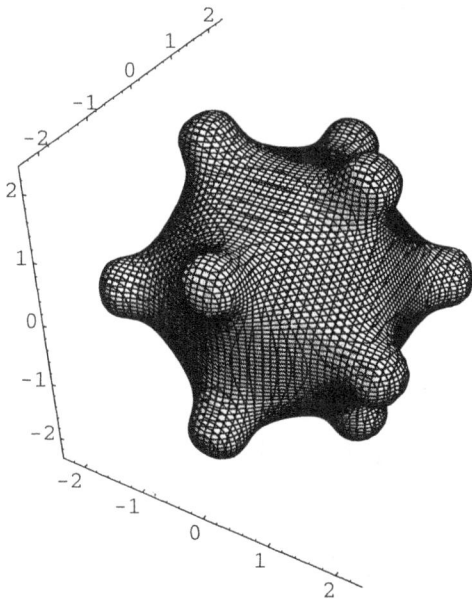

Fig 17. The opened icosahedron (fig 21, chapter 2) is here closed with a sphere.

We do the same with the tetrahedron. The equation is

$$10^{tetr} - 10^{-2tetr} + 10^{sphere} + 60 = 0 \tag{10}$$

and this gives fig 13 with its inside, the dual, in fig 14.

Again we show different split versions, in figs 15 and 16.

Finally we close the open icosahedron using the equation

$$10^{ico} - 10^{1.4dod} + 10^{sphere} = 100000 \tag{11}$$

and the result is shown in fig 17.

THE DUAL POLYHEDRON

Very similar geometry is obtained by subtracting a polyhedron from a sphere. We use the cube:

$$10^{0.1\text{sphere}} - 10^{\text{cube}} + 10000 = 0 \qquad (12)$$

This very simple equation gives a beautiful surface, shown in fig 18, with its inside, the dual, shown in figs 19 and 20.

We call this *a polyhedron* as it is built of one closed surface with polyhedral symmetry and we call it a dual of obvious reasons. By varying the contribution of the sphere and the auxiliary parameter, the isosurface constant, the shape can change considerably, as we see next. Which is the sphere minus the octahedral planes giving the dual polyhedron, the cube, and shown in fig 21. Fig 22 is a split section and the equation is below:

$$10^{0.5\text{sphere}} - 10^{\text{oct}} + 1000 = 0 \qquad (13)$$

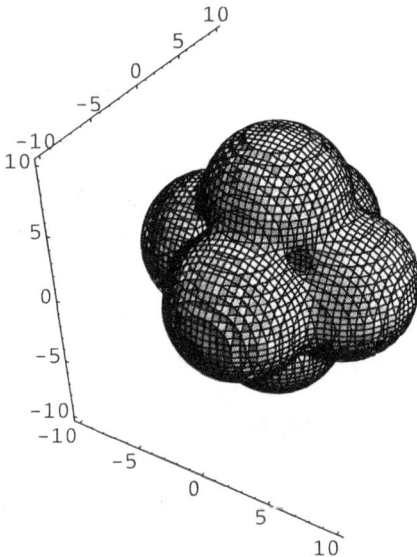

Fig 18. Subtracting a cube from a sphere gives the octahedral cluster.

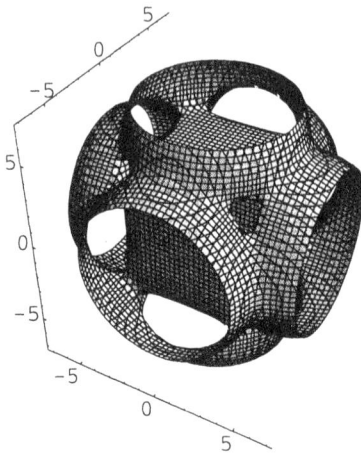

Fig 19. The inside or dual of fig 18.

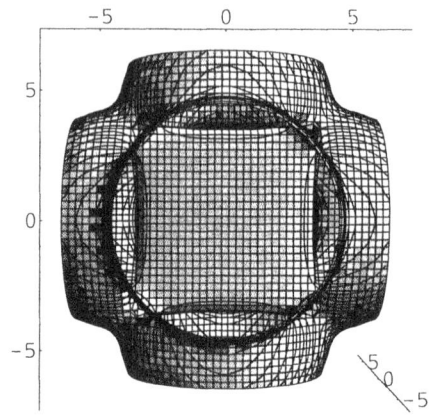

Fig 20. Another view of 19.

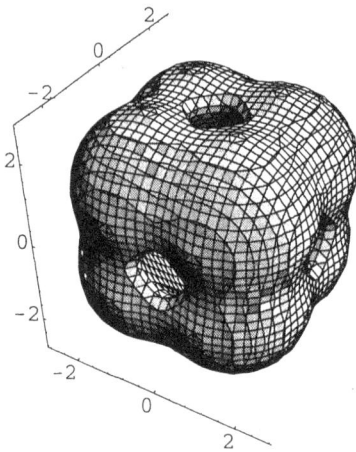

Fig 21. Subtracting an octahedron from a sphere gives the dual, cubic cluster.

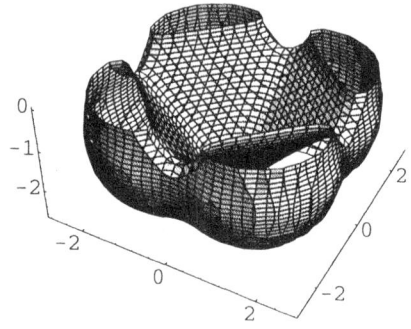

Fig 22. A split of fig 21.

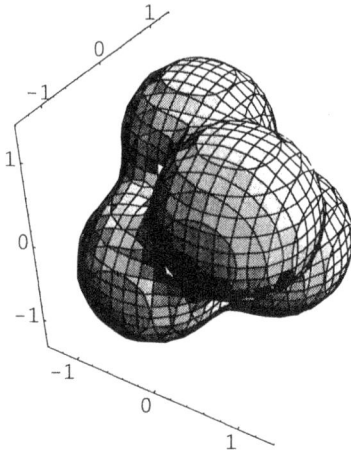

Fig 23a. Subtracting a tetrahedron from a
sphere gives its own dual.

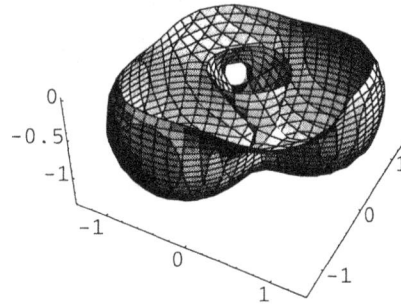

Fig 23b. A split.

Next is the tetrahedron which is its own dual of course. A very
simple equation (14) gives a beautiful surface shown in fig 23a
and in b there is a split.

$$10^{\text{sphere}} - 10^{\text{tetr}} + 5 = 0 \qquad\qquad (14)$$

We do the same with the icosahedron, using the equation

$$10^{\text{sphere}} - 10^{\text{ico}} + 10^6 = 0 \qquad\qquad (15)$$

and we get a beautiful structure, of dodecahedral shape, shown
in fig 24, with an inside of dual shape in fig 25.

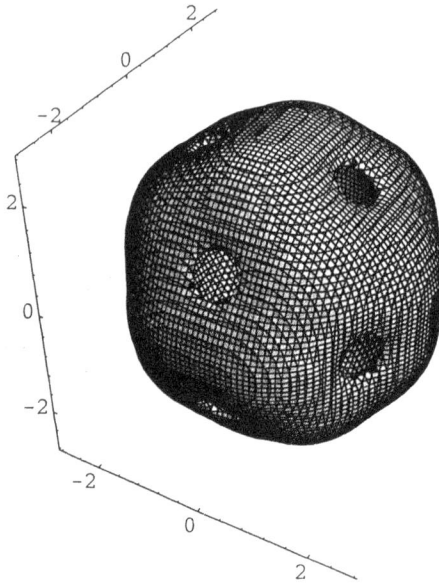

Fig 24. Subtracting an icosahedron from a sphere gives its dodecahedral dual.

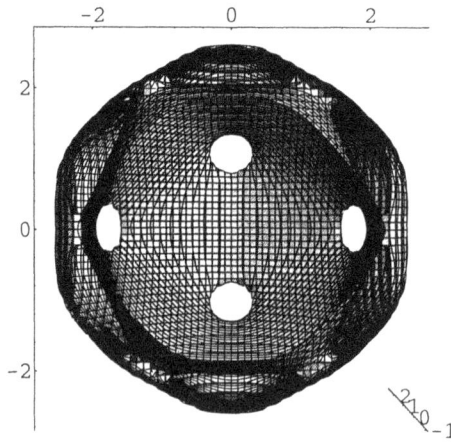

Fig 25. Split of fig 24.

CHAPTER 4

TWO DIMENSIONS

Wall paper

So far we have been in three dimensions. There are sometimes reasons for working in two. Making your own wall paper is one of them.

First we add two lines with the equation:

$$10^x + 10^y = 100 \tag{1}$$

And three:

$$10^x + 10^y + 10^{-x} = 100 \tag{2}$$

This is shown in figures 1 and 2.

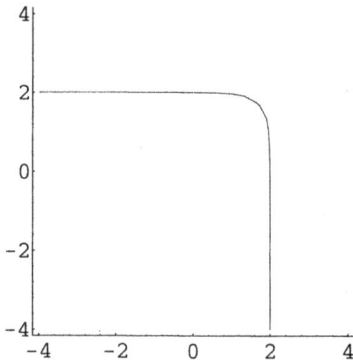

Fig 1. Two lines added on the exponential scale.

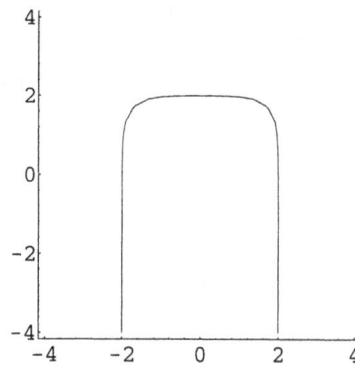

Fig 2. Three lines added on the exponential scale.

And now we do the square. The equation is:

$$10^x + 10^y + 10^{-x} + 10^{-y} = 1000 \tag{3}$$

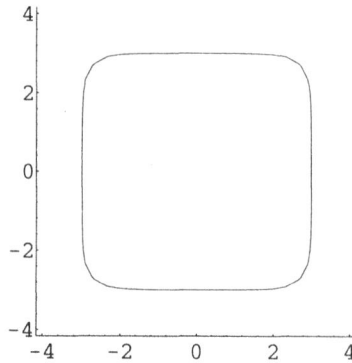

Fig 3. Four lines making a rounded square.

And the square shape is obvious in figure 3. We make the triangle and the hexagon later. First we do structures which have repeating patterns. We find such in chemical structures - in crystals. Repeating patterns are obtained from algebraic curves by replacing x and y by trigonometric functions of x and y. We start with the equation

$$10^{\cos 2\pi x} + 10^{\cos 2\pi y} = \text{const} \tag{4}$$

and for const=1 we have a splendid structure in figure 4 (or wall paper).

What we do now is a scheme that will be repeated in this chapter. We make the atoms (we call these things atoms because that's the way we think, chemists as we are) approach each other by changing the constant, they interact and a new pattern is formed.

The constants are 9, 10.1, 10.3, for figures 5a, b, and c. Fig 5d and e, the wall paper, has the same constant, 10.5. All with equation (4).

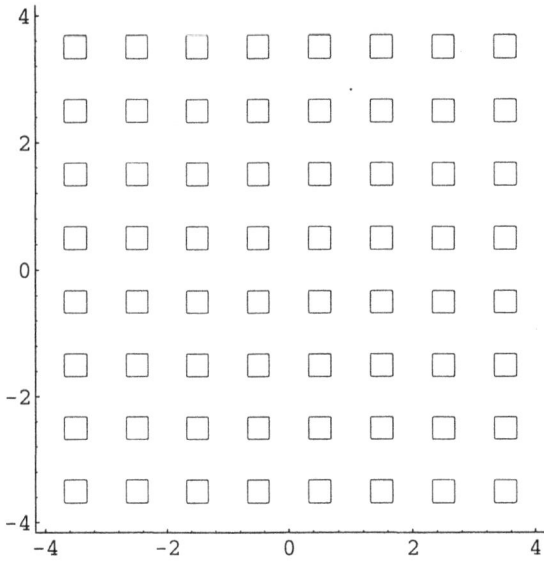

Fig 4. Cosine modulation of two lines making wall paper.

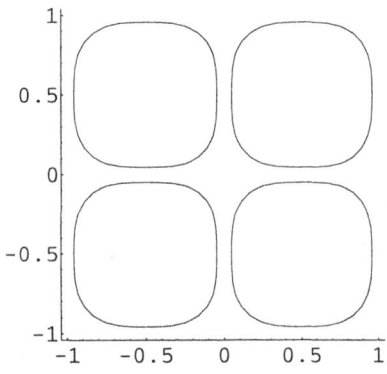

Fig 5a. Eq. (4) with const=9.

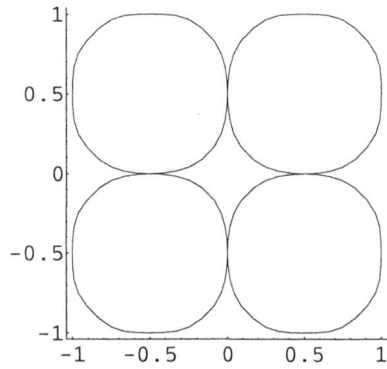

Fig 5b. Eq. (4) with const=10.1.

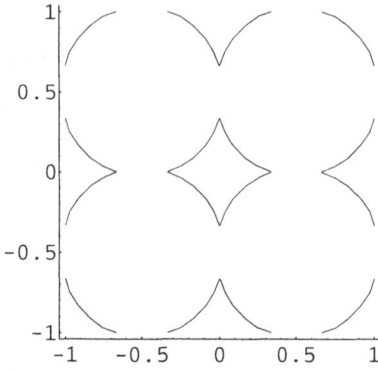

Fig 5c. Eq. (4) with const=10.3.

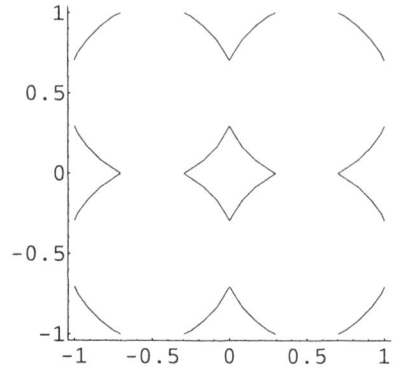

Fig 5d. Eq. (4) with const=10.5.

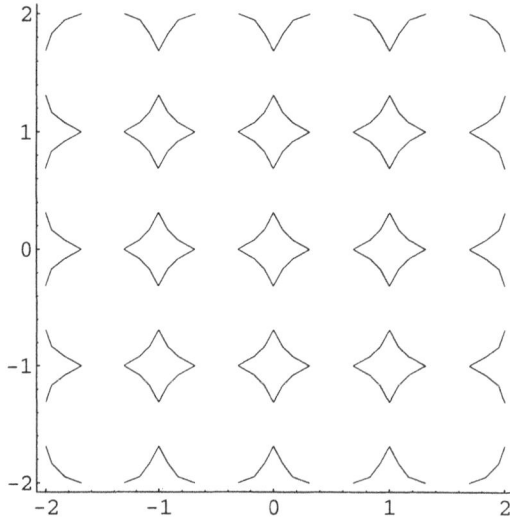

Fig 5e. Larger part of the structure in d.

Scaling the exponent or changing the base results in sharper or rounder atoms. For example a lower base, like *e*, results in rounder atoms. With constants 2, 2.9 and 3.1 in eq. (5) we have figures 6a, b, and c.

$$e^{\cos 2\pi x} + e^{\cos 2\pi y} = const \qquad (5)$$

The general relationship between the base and scaling of the exponent is:

$$B^x = 10^{\log B \cdot x} \qquad\qquad (6)$$

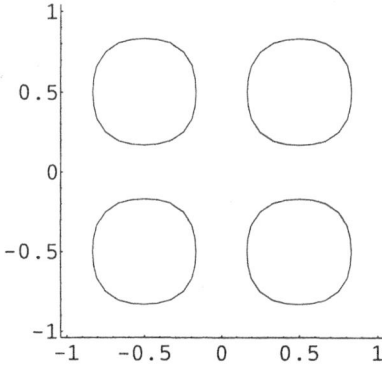

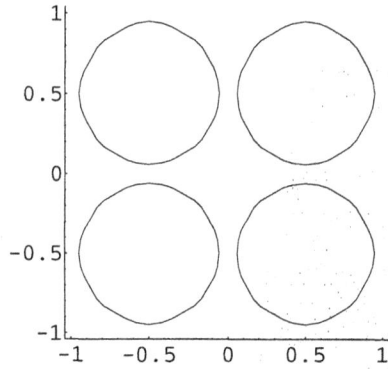

Fig 6a. With e as base the 'atoms' are rounder. Eq. (5) with const=2.0.

Fig 6b. Eq. (5) with const=2.9.

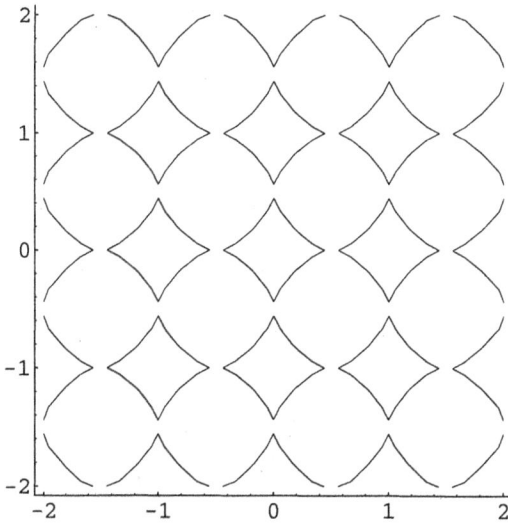

Fig 6c. Eq. (5) with const=3.1.

Surely one could now do the seventeen two-dimensional groups of isometries, or the ornaments of Alhambra. Each of them condensed into one simple equation. We have not done that yet - but we have done some more wall paper. We report below just giving equations and structures. Equation (7) is illustrated in fig 7, and equation (8) in fig 8.

$$10^{\cos 2\pi x} - 10^{\cos 2\pi y} = 0 \tag{7}$$

$$10^{\cos 2\pi x} - 10^{\cos \pi y} = 0 \tag{8}$$

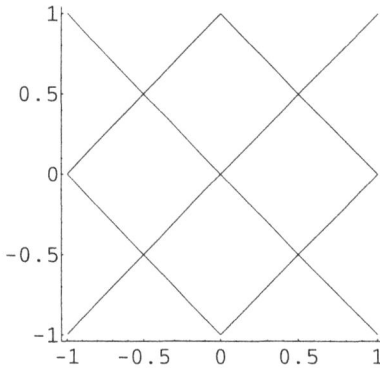

Fig 7. Eq. $10^{\cos 2\pi x} - 10^{\cos 2\pi y} = 0$

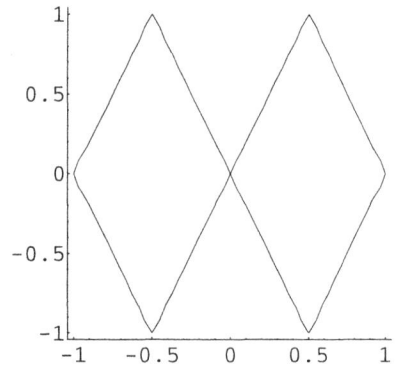

Fig 8. Eq. $10^{\cos 2\pi x} - 10^{\cos \pi y} = 0$

The function (9) gives us the nice pattern below in fig 9.

$$10^{\cos 2\pi(x-2y)} + 10^{\cos 2\pi(x+2y)} = 8.4 \tag{9}$$

The function (10) gives figures 10 and 11 for constants 4 and 11.

$$10^{\cos 2\pi(x-y)} + 10^{\cos 2\pi(x+y)} +$$
$$10^{\cos 2\pi 2x} + 10^{\cos 2\pi 2y} = const \tag{10}$$

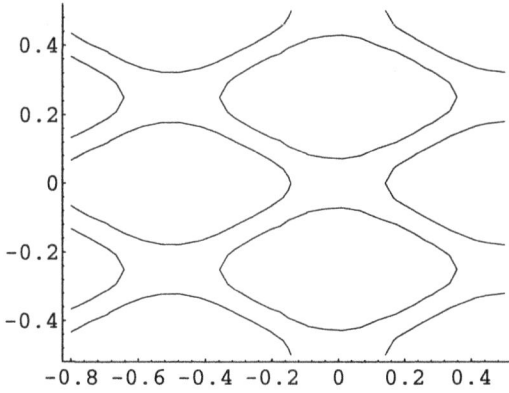

Fig 9. Eq. $10^{\cos 2\pi(x-2y)} + 10^{\cos 2\pi(x+2y)} = 8.4$

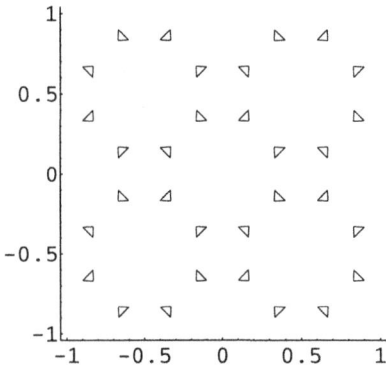

Fig 10. Eq. $10^{\cos 2\pi(x-y)} + 10^{\cos 2\pi(x+y)} + 10^{\cos 2\pi 2x} + 10^{\cos 2\pi 2y} = 4$

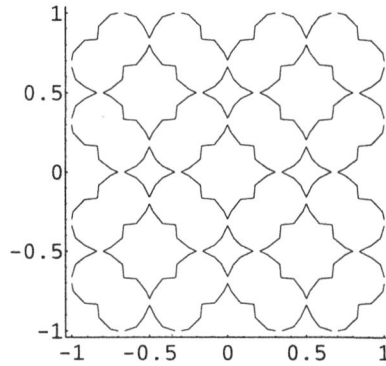

Fig 11. Eq. $10^{\cos 2\pi(x-y)} + 10^{\cos 2\pi(x+y)} + 10^{\cos 2\pi 2x} + 10^{\cos 2\pi 2y} = 11$

The zero difference between terms (equation 11) give a beautiful net in fig 12

$$10^{\cos 2\pi(x-y)} - 10^{\cos 2\pi(x+y)} +$$
$$10^{\cos 2\pi 2x} - 10^{\cos 2\pi 2y} = 0 \tag{11}$$

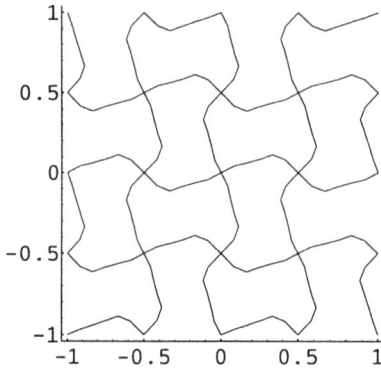

Fig 12. Eq.
$$10^{\cos 2\pi(x-y)} - 10^{\cos 2\pi(x+y)} + 10^{\cos 2\pi 2x} - 10^{\cos 2\pi 2y} = 0$$

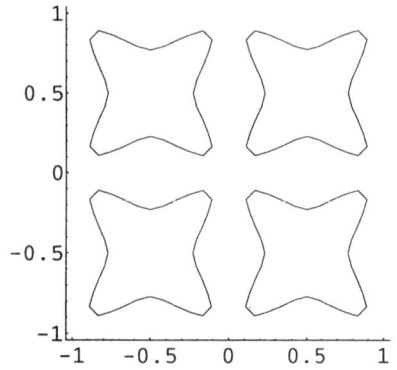

Fig 13. Eq.
$$10^{\cos 2\pi(x-y)} + 10^{\cos 2\pi(x+y)} - 10^{\cos 2\pi 2x} - 10^{\cos 2\pi 2y} = 0$$

The other zero difference (equation 12) gives the formidable pattern of fig 13.

$$10^{\cos 2\pi(x-y)} + 10^{\cos 2\pi(x+y)} -$$
$$10^{\cos 2\pi 2x} - 10^{\cos 2\pi 2y} = 0 \qquad (12)$$

Now we make the triangle and the hexagon with the equations:

$$100^{x+0.577y} + 100^{-1.15y} + 100^{0.577y-x} = 100 \qquad (13)$$

$$10^{x+0.577y} + 10^{-1.15y} + 10^{0.577y-x} +$$
$$10^{-(x+0.577y)} + 10^{1.15y} + 10^{-(0.577y-x)} = 1400000 \qquad (14)$$

$$10^{\cos 2\pi(x+0.577y)} + 10^{\cos 2\pi(-1.15y)} +$$
$$10^{\cos 2\pi(0.577y-x)} + 10^{-\cos 2\pi(x+0.577y)} + \qquad (15)$$
$$10^{-\cos 2\pi(1.15y)} + 10^{-\cos 2\pi(0.577y-x)} = 14.25$$

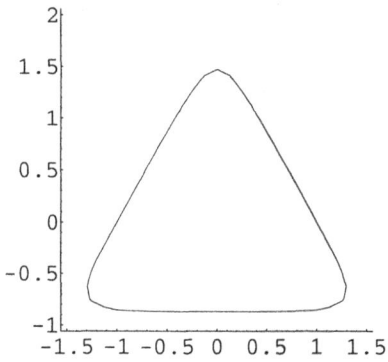

Fig 14. The triangle of eq. (13).

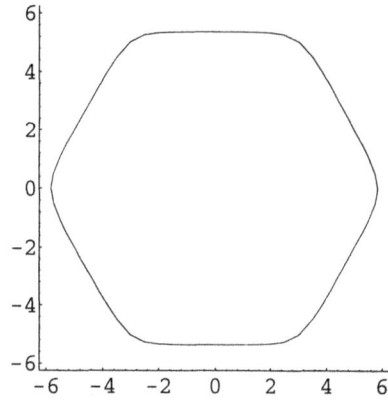

Fig 15. Equation (14) giving the hexagon.

We have chosen to modulate the hexagon to produce wall paper from equation (15) and in fig 16 we see the result in form of regularly repeated hexagons. By varying the constant, 16 and 12, we get the two kinds of hexagonal nets, as seen in figures 17 and 18.

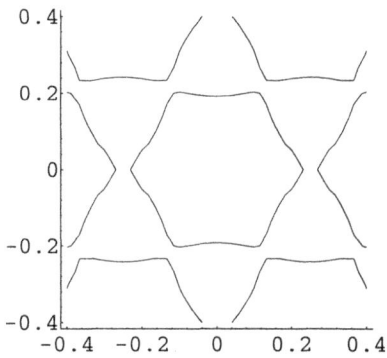

Fig 16. Eq. (15) with a const of 14.25.

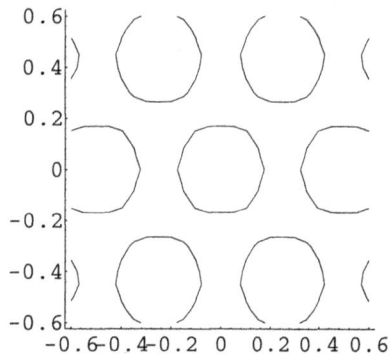

Fig 17. Eq. (15) with a const of 16.

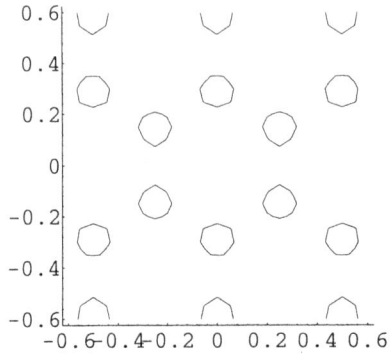

Fig 18. Eq. (15) with a const of 12.

CHAPTER 5

THREE DIMENSIONAL STRUCTURES

As an introduction we use direct trigonometric modulation of face vectors in three dimensions and derive chemical structures. The so-called minimal surfaces - of great importance in mathematics - seem to have an interest to science limited to soap water. Surfaces related to these, and many others, will show up as special cases, in our hurdle after structures and the related nets.

To show the principles we start with the column structure from chapt. 1. We only need two perpendicular face vectors, as the cosine modulation (sine is also OK) will do the rest.

We use the equation

$$10^{\cos(2\pi x)} + 10^{\cos(2\pi y)} = \text{const} \qquad (1)$$

and get for const=7.0 our first structure in fig 1.

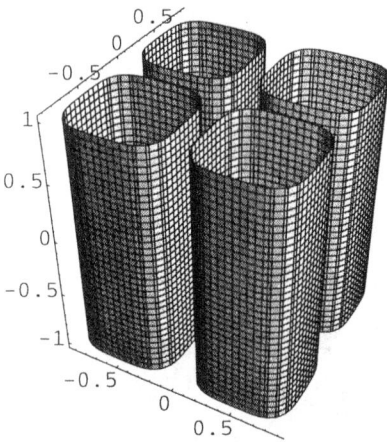

Fig 1. Modulation of the column structure of Fig 3 chapter 1.

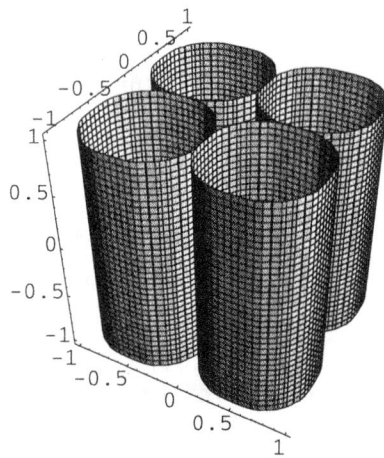

Fig 2. Change of const. to 10.1.

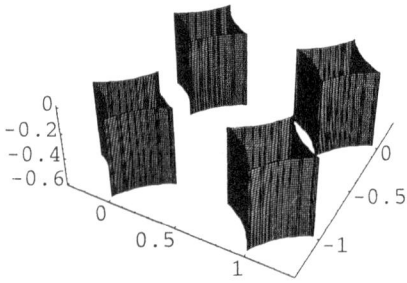

Fig 3. Change of const. to 10.5.   **Fig 4.** Projection of fig 3.

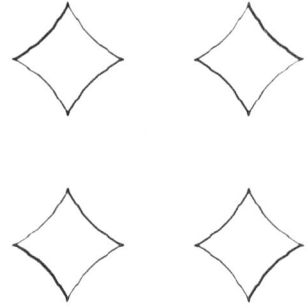

At a constant of 10.1 the columns touch, and at 10.5 the structure has inverted to new columns, shown in figs 2 and 3.

Fig 4 is a thin slice of fig 3 projected along c. This is a 3D calculation exactly analogous to the 2D ones in the wall paper case, see equation (9) and figures 5a-e.

To complete we also give the simple function

$$\cos(2\pi x) + \cos(2\pi y) = 0 \tag{2}$$

being the first term of the power expansion from our exponential scale (fig 5). The result is intersecting planes.

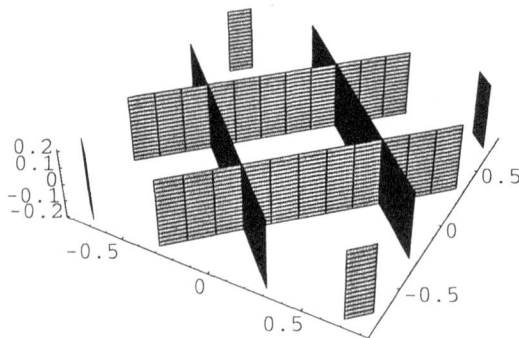

Fig 5. Eq. (2) gives intersecting planes.

Nodal surfaces

So far we have done simple trigonometric modulation to get repetition. We used to perform this also on the polyhedral equations to arrive at a mathematical descriptions of crystal structures. We have recently found a different approach, which we prefer, and show below.

One of the most important functions in mathematics is the exponential one:

$$e^x + e^y + e^z - \text{const} = 0 \tag{3}$$

which we know is a cube corner in 3D.

We start to use complex numbers. If you are not used to such things listen here. First of all - we use them because they are useful. Writing this book we wanted to get the nodal surfaces in a simple way. That is why we started to use i. Doing this we realised we had a way - and a simple way - of deriving nodal surfaces. We will show you how it happened. You might like the use of i !

i is an imaginary number, it does not exist for us. Similar to minus one apple. i is $\sqrt{-1}$.

If we accept

$$e^{ix} = \cos x + i \sin x \tag{4}$$

we also understand the famous definitions of the cyclic functions:

$$\sin x = \frac{1}{2} i (e^{ix} - e^{-ix}) \tag{5}$$

$$\cos x = \frac{1}{2} (e^{ix} + e^{-ix}) \tag{6}$$

Using e^{ix} means the real part is cosx and the imaginary part is sinx. Or

$$\text{Re}[e^{ix}] = \cos x \tag{7}$$

and

$$Im[e^{ix}] = \sin x \qquad (8)$$

But the general thing to use is *the complex exponential*:

$$e^{\pi ix} + e^{\pi iy} + e^{\pi iz} \qquad (9)$$

We write

$$e^{ix} = \cos x + i \sin x \qquad (10)$$

and the real part of the complex exponential is

$$Re[e^{\pi ix} + e^{\pi iy} + e^{\pi iz}] = \cos \pi x + \cos \pi y + \cos \pi z \qquad (11)$$

(Did you notice we got the P-nodal surface!)

First, we simplify equation (4) with the famous formula

$$e^{\pi i} = -1 \qquad (12)$$

and we get

$$i^{2x} + i^{2y} + i^{2z} \qquad (13)$$

which is identical with (9).

We show you the geometrical meaning of three equations:

$$i^{2x} + i^{2y} + i^{2z} = 0 \qquad (14)$$

$$i^{2x} + i^{2y} + i^{2z} = 1 \qquad (15)$$

$$i^{2x} + i^{2y} + i^{2z} = 2 \qquad (16)$$

Equation Re[(14)] is the P - nodal surface which is shown in fig 6a. Equations Re[(15)] and Re[(16)] are shown in figures 6b and 6c.

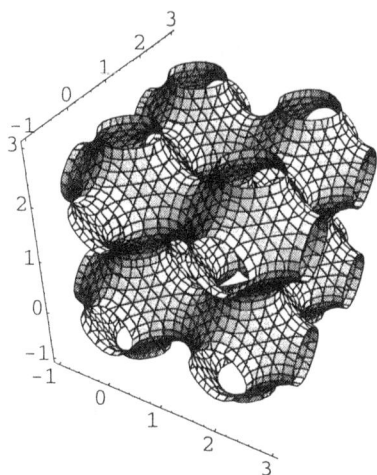

Fig 6a. Re part of the eq. (14) gives the P-nodal surface.

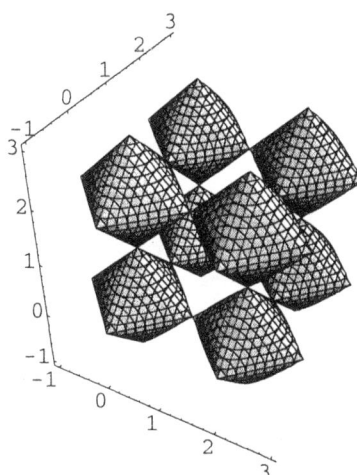

Fig 6b. Re parts of the eq. (15) gives the ReO₃ structure type.

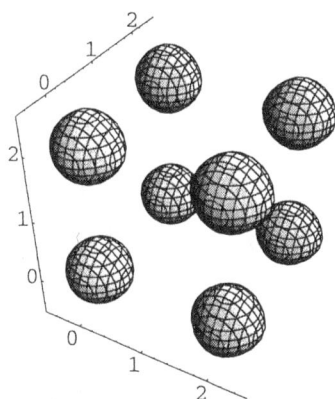

Fig 6c. Re parts of the eq. (16) gives pc packing of bodies.

These figures represent the two simplest structures ever - the primitive cubic packing of atoms and the so called ReO_3 structure. The ReO_3 structure has metal atoms in the centre of the oxygen octahedra.

To some surprise we realise that the famous complex exponential in three dimensions is the equally famous (almost) P - nodal surface in geometry! Perhaps you now think - stop, this is enough, periodicity (and symmetry) and the complex exponential are synonyms - the rest is trivial. But it is not. We are trying to describe crystal structures, in the way the logarithmic spiral describes Nautilus. And the concept of crystal structures is immensely richer than that of symmetry. We need to continue with the more advanced trigonometry on the exponential scale. That is what the main part of the rest of this book deals with.

So we just did i^{2x}. The next obvious combination of the space variables gives the expression:

$$i^{x+y} + i^{y+z} + i^{z+x} + i^{x-y} + i^{y-z} + i^{z-x} \qquad (17)$$

For its Im respectively Re part it is identical with the Gyroid and IWP nodal surface. With this we mean the following equation for the IWP nodal surface which is identical with the Fermi surface in reciprocal space for *ccp* [8] and shown in fig 7a:

$$\mathrm{Re}[i^{x+y} + i^{y+z} + i^{x+z} + i^{x-y} + i^{y-z} + i^{-x+z}] = -1 \qquad (18)$$

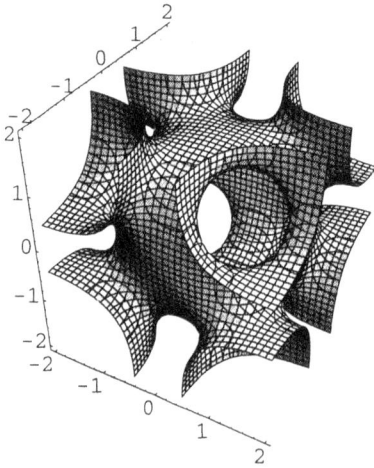

Fig 7a. Eq. (18) gives the IWP type surface.

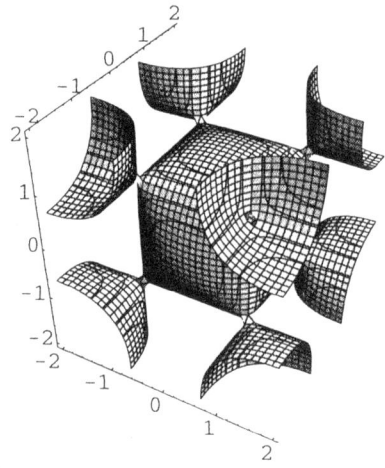

Fig 7b. A constant of 0 as in eq. (19) results in corner sharing cubes.

The normal nodal equation

$$Re[i^{x+y} + i^{y+z} + i^{x+z} + i^{x-y} + i^{y-z} + i^{-x+z}] = 0 \quad (19)$$

is just cubes sharing corners as seen in fig 7b.

Its Im part with constant of 0 gives the gyroid nodal surface shown in fig 7c.

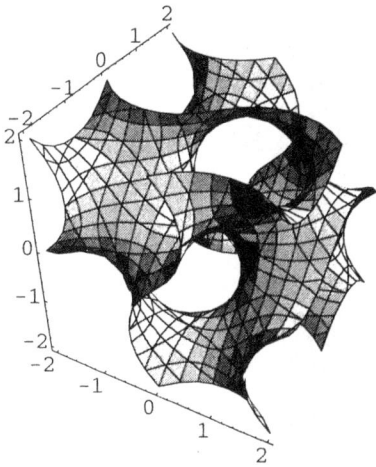

Fig 7c. The gyroid as a result of the Im part of eq. (19).

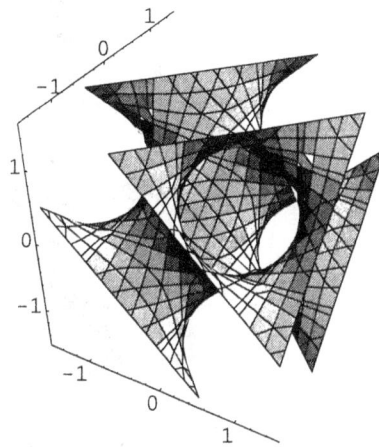

Fig 7d. Adding the Re and Im parts of eq. (20) gives the D-type surface.

Changing constant gives of course the **bcc** structure. Finally the equation

$$i^{-x-y-z} + i^{x-y+z} + i^{x+y-z} + i^{-x+y+z} = 0 \quad (20)$$

gives the D nodal surface adding the Re and Im parts (fig 7d), and the diamond structure by adding a constant. The Re or Im parts on their own give intersecting planes.

We can mix equations (13) and (17) to (21) below, and get the very remarkable surface in fig. 8. We shall analyse this one later.

$$i^{x+y} + i^{y+z} + i^{x+z} + i^{x-y} + i^{y-z} + i^{-x+z} +$$
$$i^{2x} + i^{2y} + i^{2z} = 0 \tag{21}$$

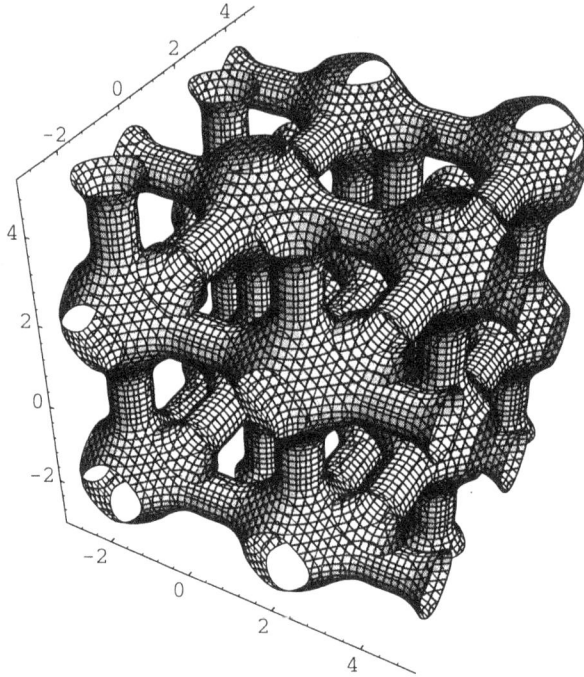

Fig 8a. Adding the two equations (13) and (17) gives two, identical and interpenetrating surfaces.

Similarly we construct:

$$i^{-x+y+z} + i^{x+y-z} + i^{x-y+z} + i^{-x-y-z} +$$
$$i^{x} + i^{y} + i^{z} = 0 \tag{22}$$

and get a simple answer in form of the Neovius nodal surface in
fig 8b.

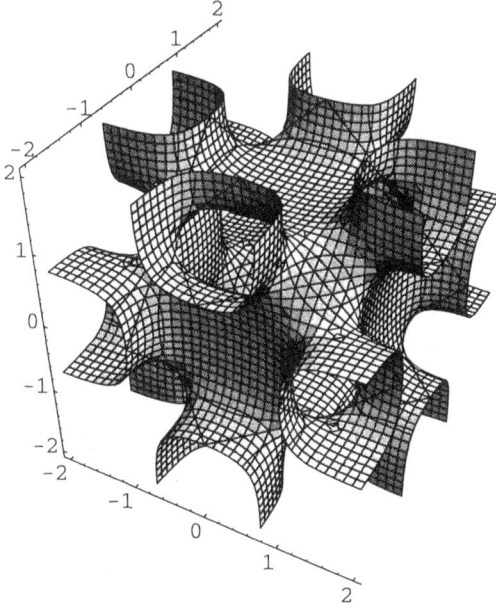

Fig 8b. Adding the two equations (13) and (20) gives a
surface of Neovius type.

We conjecture that the rest of the nodal surfaces, as published by
Nesper and von Schnering [15], should exist as various
combinations of the fundamental equations given above.

We have shown the link between the exponential scale and the
well known nodal surfaces. As well as a proper way to derive
these surfaces.

The crystal structures described in this book will be derived
from the functions below:

$$i^{2x} \tag{23}$$

$$e^{i^{2x}} \tag{24}$$

$$e^{i^{2x}} e^{i^{-2x}} \tag{25}$$

$$e^{-ii^{2x}} e^{ii^{-2x}} \tag{26}$$

(25) and (26) are the real and imaginary parts of

$$e^{\cos \pi x} + ie^{\sin \pi x} \tag{27}$$

From these complex exponentials in space stems the fundamental cubic symmetries, and their periodic structures. Those are the primitive, the body centred, the face centred, and the diamond symmetries.

The functions (24), and (25) or (26) will be used in this study - we start with (27) being the simpler. (25) and (26) are just an application of Euler's formulas to the exponential scale.
The natural base e is by no means a natural base for this study. We will show the advantages of a higher base, using 10 or 100, and compare that with e ($100^x = 10^{2x} = e^{\ln(100)x}$).

$B \cos x$

We start with Re(27):

$$e^{\cos 2\pi x} + e^{\cos 2\pi y} + e^{\cos 2\pi z} = \text{const} \tag{28}$$

In fig 9a we see the primitive cubic packing (*pc*) of atoms at a constant of 3.

Increasing constant means increasing sizes of the bodies - exactly the same mechanism as blowing a balloon. We can also say the bodies approach each other with augmenting constant and finally touch each other at a constant of 3.454 as shown in fig 9b. We regard this variation as a change of time, as an auxiliary parameter - we call it *mathematical dynamics*. If the bodies are atoms this provides us with a picture of force - all the atoms move simultaneously under the influence of mutual attraction and finally form a surface. When interacting, the spheres, representing atoms, develop catenoids, or bonds, between themselves and via monkey saddles finally form a periodic surface of nodal type. This is shown in fig 9c where catenoids have started to develop between spheres at a constant of 3.458. At a constant of 4 a surface of P nodal type is fully developed as shown in fig 10. It has been shown by von Schnering and Nesper[4] that electronic structure often is explained by such a surface.

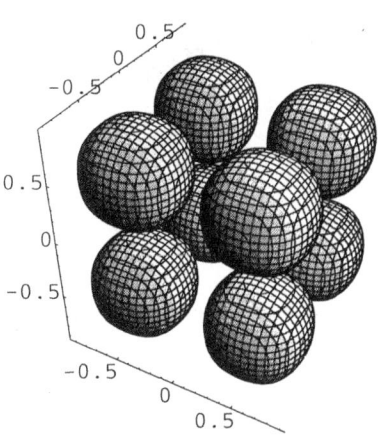

Fig 9a. Eq. (28) with const=3.0 shows the primitive cubic packing of bodies.

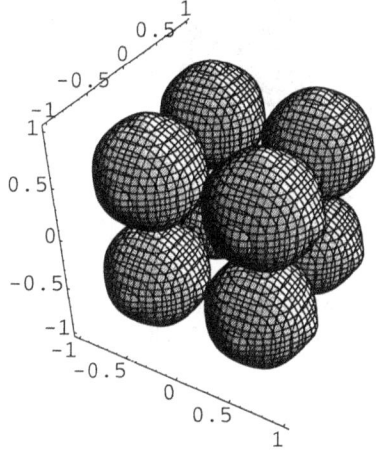

Fig 9b. Changing the constant to 3.454 or $e+2e^{-1}$ increases the size of the bodies.

The exact expression for atoms touching each other as in fig. 9b is:

$$e^{\cos 2\pi x} + e^{\cos 2\pi y} + e^{\cos 2\pi z} = e + 2e^{-1} \qquad (29)$$

and the ReO_3 similar structure in fig 11 has the exact formula:

$$e^{\cos 2\pi x} + e^{\cos 2\pi y} + e^{\cos 2\pi z} = 2e + e^{-1} \qquad (30)$$

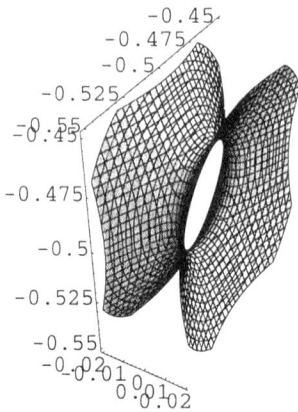

Fig 9c. A cateniod between
bodies at a constant of 3.458.

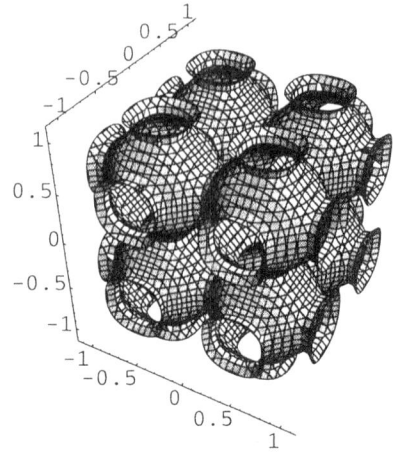

Fig 10. Equation (28) with const = 4.0.

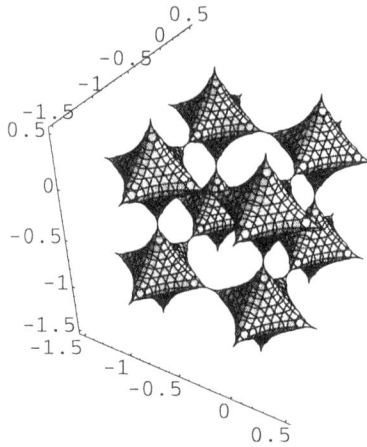

Fig 11. Equation (28) with const = $2e+e^{-1}$.

Changing the base to 100 makes atoms square and we show
from equation (27) two figures to demonstrate this, fig 12 with a
constant of 2, and fig 13 with the constant 20 in equation (31).

With this higher base, the cubic overall symmetry is also heavily influencing the bodies in the structure.

$$100^{\cos 2\pi x} + 100^{\cos 2\pi y} + 100^{\cos 2\pi z} = const \qquad (31)$$

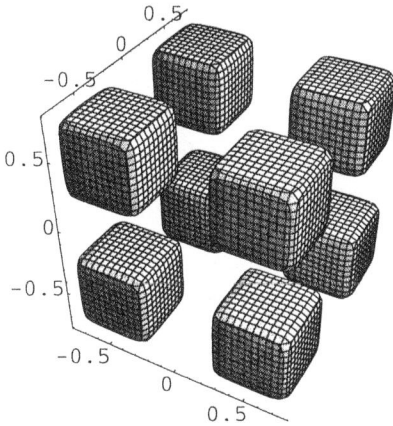

Fig 12. A change of base to 100 as in eq. (31) and a constant of 2.0 gives square bodies.

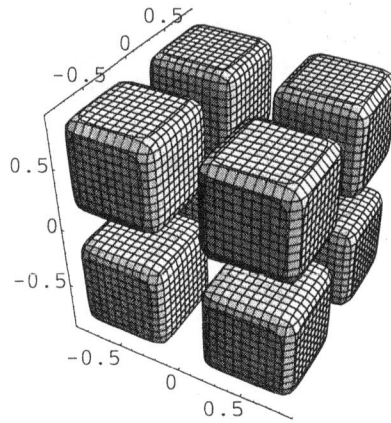

Fig 13. As in fig 12 but const = 20.0.

At a constant of $10^2 + 2 \cdot 10^{-2}$ the cubes touch (fig 14) and we may say we have the CsCl structure (Cs in the centre of Cl cubes).

For a constant of 170 we get fig 15, a splendid net for primitive structures like ReO_3. Here we clearly see the difference using different bases.

We are now better off for analysing the surface in fig 8a. It is two nets of ReO_3 type of fig 15, interpenetrating into each other. Remarkably enough there is a correspondent in nature - the structure of Nb_6F_{15} [23], once a model for the application of minimal surfaces and curvature to chemistry[10]. The two nets are really parallel surfaces to a P nodal surface situated between them.

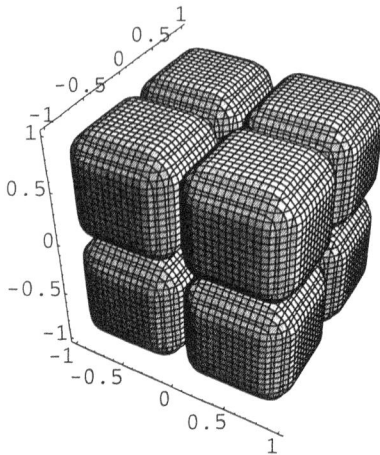

Fig 14. As in fig 12 but const = $10^2 + 2 \cdot 10^{-2}$, to give the CsCl structure.

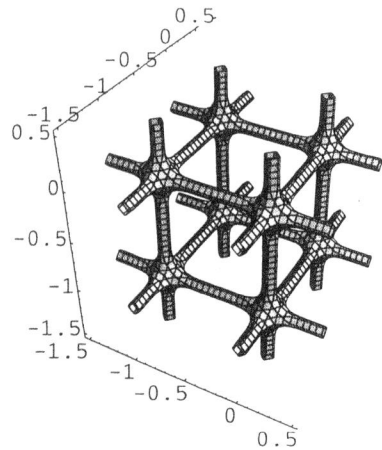

Fig 15. As in fig 12 but const = 170 , to give the ReO_3-net.

$B^{cos(x+y)}$

The variation in space of this exponential is given in (32):

$$e^{\cos 2\pi(x+y)} + e^{\cos 2\pi(x-y)} + e^{\cos 2\pi(y+z)} +$$
$$e^{\cos 2\pi(y-z)} + e^{\cos 2\pi(x+z)} + e^{\cos 2\pi(x-z)} = \text{const} \tag{32}$$

The structure seen in fig 16 a and b with a constant of 5.5 in equation (32). The bodies represent the vertices of a truncated octahedron which is the domain of an atom in **bcc**. Such an atom has eight nearest neighbours and six at a slightly greater distance. The eight hexagonal and six cubic faces of the truncated octahedron correspond to these two sets of distances in **bcc**. This polyhedron is space filling and its corners correspond to the Al and Si atoms in the chemically fundamental structure of Sodalite.

We make the spheres fuse together by a constant of 6 and get a structure which clearly is that of Sodalite - the simplest and most fundamental of all zeolite structures, in fig 17. The metal atoms are here situated inside the tetrahedral voids, and the oxygen in the catenoids bridging.

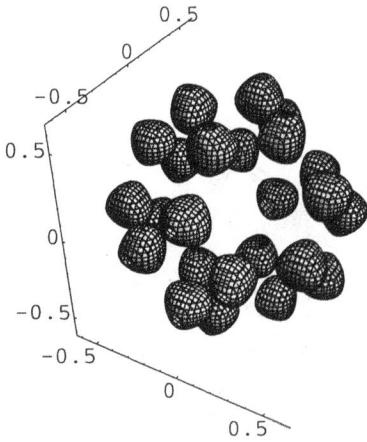

Fig 16a. The equation (32) with const=5.5 shows bodies as the metal atoms in Sodalite.

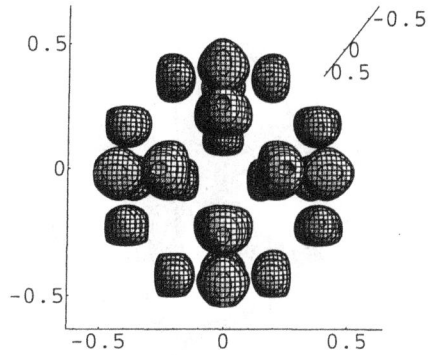

Fig 16b. Different view of 16a.

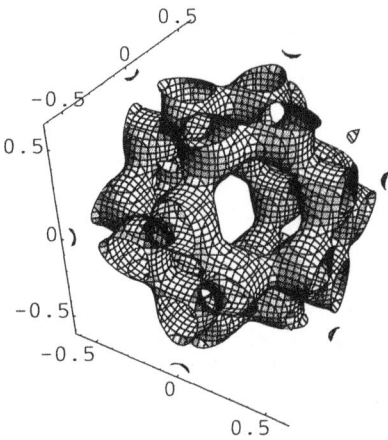

Fig 17. As in Fig 16 but const=6 gives the net of Sodalite.

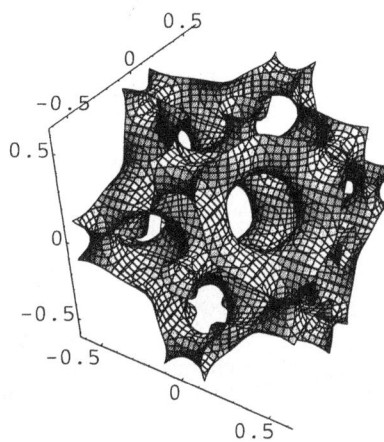

Fig 18. As in Fig 16 but const=6.5 gives the surface of Sodalite.

The surface character is clear in fig 18 with a constant of 6.5 but we transform it to a net by shifting the base to 100 in fig 19. The constant is now 185 and the *bcc* symmetry is beautifully clear.

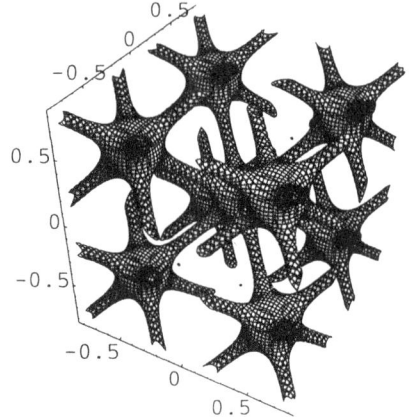

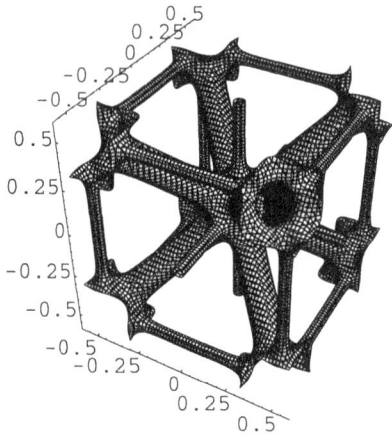

Fig 19. Changing the base to 100 in the equation of fig 16, and the const to 185, gives the **bcc** net.

Fig 20. With a base of 10 and a const of 28 a simpler **bcc** net is obtained.

Changing the base to reduce the net character to 10 and with a constant of 28 we get the equation

$$10^{\cos 2\pi(x+y)} + 10^{\cos 2\pi(x-y)} + 10^{\cos 2\pi(y+z)} +$$
$$10^{\cos 2\pi(y-z)} + 10^{\cos 2\pi(x+z)} + 10^{\cos 2\pi(x-z)} = 28 \tag{33}$$

which gives the beautiful net of figure 20.

This remarkable surface in fig 20 represents of course any chemical structure of corner sharing cubes. $NiHg_4$ is one, and Ni is then in the centres and Hg in the corners of the cube.

Going back to the base of *e* in order to get rounder atoms we use equation (34) to produce *bcc* of atoms as seen in figure 21.

$$e^{\cos 2\pi(x+y)} + e^{\cos 2\pi(x-y)} + e^{\cos 2\pi(y+z)} +$$
$$e^{\cos 2\pi(y-z)} + e^{\cos 2\pi(x+z)} + e^{\cos 2\pi(x-z)} = 11 \tag{34}$$

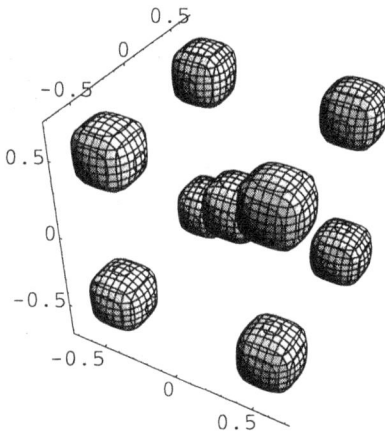

Fig 21. Equation as in Fig 16 and a const of 11 gives 'rounder atoms' to the **bcc** structure.

$B \, cos(x+y+z)$

The permutations in space are as in the equation below which is also used in the first calculation.

$$10^{\cos 2\pi(x+y-z)} + 10^{\cos 2\pi(x-y+z)} +$$
$$10^{\cos 2\pi(-x+y+z)} + 10^{\cos 2\pi(-x-y-z)} = 2 \tag{35}$$

In fig 22 the Kepler stella octangula arrangement of spheres is the geometrical meaning of the contour plot of function. This is also the unit cell of *fcc*, or the structure of cubic close packing of bodies.

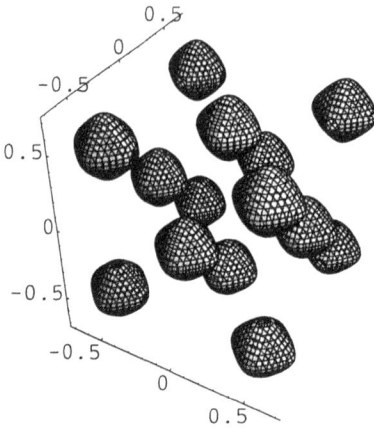

Fig 22. The equation (35).

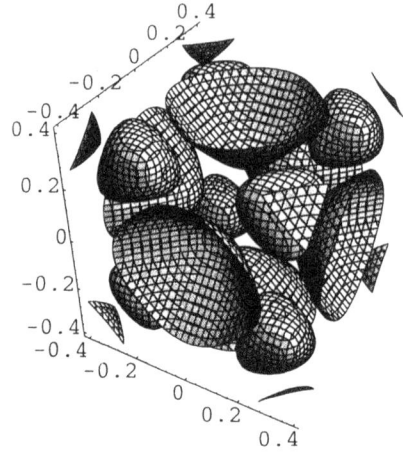

Fig 23. Equation as in fig 22 but at a const of 10.0 small tetrahedra show up between the atoms.

Instead of direct fusion between bodies when the constant is increased, small rounded tetrahedra are formed during the dynamics, as shown in fig. 23. These are situated in the tetrahedral interstices of the close packing. Chemically this corresponds to the hydrogens in TiH_2, the larger spheres being the titanium. This structure belongs to the CaF_2 family.

At a constant of 11.5 catenoidic fusion finally occurs between the small and the large bodies in a tetrahedral manner (fig 24).

This structure quickly gets a net character making the surface hard to analyse as seen in fig 25 (const 13), and also in fig 26 (const 14). The polyhedron behind this net is the cube octahedron, the arch polyhedron for *fcc* and fig 27 gives a more complete net now at a constant of 17. To reduce the net character we switch to a lower base and the equation (36) below gives fig 28.

$$e^{\cos 2\pi(x+y-z)} + e^{\cos 2\pi(x-y+z)} +$$
$$e^{\cos 2\pi(-x+y+z)} + e^{\cos 2\pi(-x-y-z)} = 5.2 \tag{36}$$

It is now clear that this surface belongs to the FRD- type.

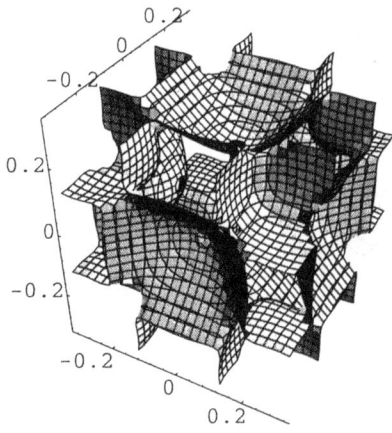

Fig 24. Const = 11.5.

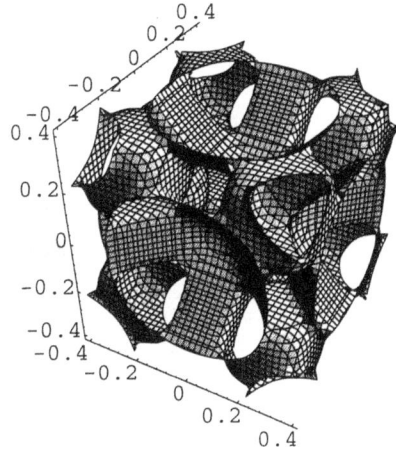

Fig 25. Const = 13.0.

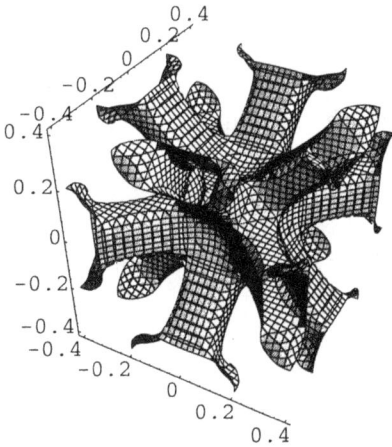

Fig 26. Const = 14.0.

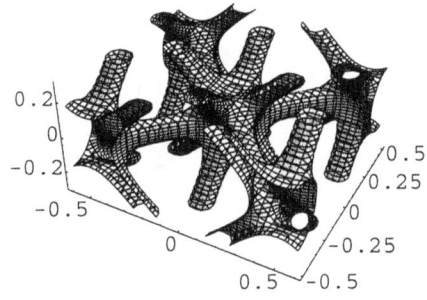

Fig 27. Const = 17.0.

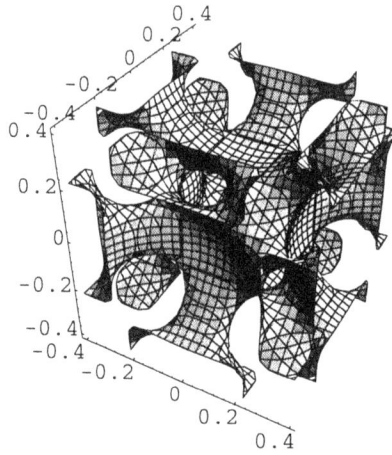

Fig 28. Equation as in fig 22 but using
the lower base of e and a const of 5.2. The
surface is of the FRD type.

$$B \, sin(x+y+z) \quad + B \, cos(x+y+z)$$

In order to obtain the diamond symmetry we need to use both
sin and cos modulation of the tetrahedral symmetry, according to
the equation below.

$$e^{\cos 2\pi(x+y-z)} + e^{\cos 2\pi(x-y+z)} +$$
$$e^{\cos 2\pi(-x+y+z)} + e^{\cos 2\pi(-x-y-z)} +$$
$$e^{\sin 2\pi(x+y-z)} + e^{\sin 2\pi(x-y+z)} + \tag{37}$$
$$e^{\sin 2\pi(-x+y+z)} + e^{\sin 2\pi(-x-y-z)} = C$$

The dynamics of this function behaves similar to the earlier cases
studied. At a low constant diamond structure "atoms" form, and
these then fuse together to a polyhedral network of corner
connected tetrahedra with increasing constant. This then turns
into a D-type surface, to finally reverse on the other side. In the
final atomic structure the atoms have the same packing as the
original, but are situated on the other side of the surface. The

dynamics are illustrated in fig 29 in the area of -0.6 to +0.6 in x, y and z, and with the isosurface constant indicated for each step.

The fusing of atoms gives a feeling of force involved in the functions, and to somehow visualise this force it is possible to study the gradient of the function, as pointed out by R. Nesper. Therefore the isosurfaces in the series of the diamond dynamics in fig 29 are coloured by the value of the gradient of the function. Purple indicates the lowest gradient values, and white the highest, according to the colourmap given in the figure. We see that fusing occurs at points of high curvature and low gradient values. The flatpoints show lowest degree of fusing, with maximum gradient values. The trend of the gradient is also valid in the other structures in the exponential scale.

With the base 10 instead of e, and with the constant 13, a small sphere of interaction is formed between the larger ones, as seen in fig 30. The position of this is that of the sp^3 pair of electrons in diamond, or of the oxygens in cristobalite, a well-known form of SiO_2.

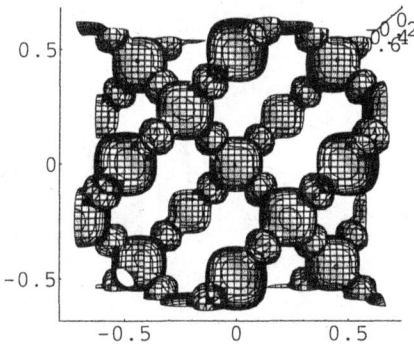

Fig 30. Equation as in fig 29, but with the base 10 and a const of 13.0, gives the diamond structure with electrons.

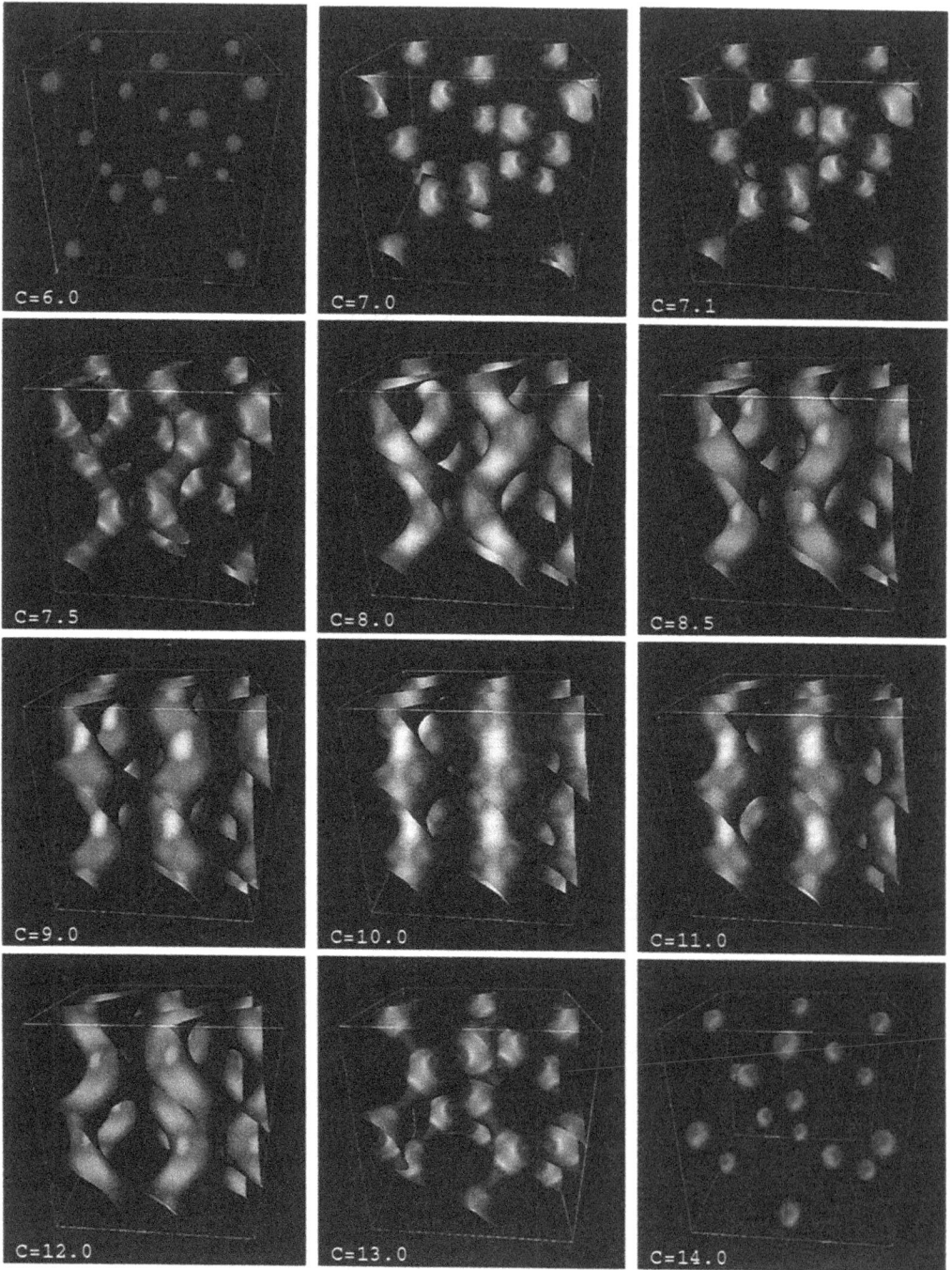

Fig 29. Eq. (37) coloured by its gradient with the colourmap:

$$e^{i^{2x}} = e^{\cos \pi x} \cos(\sin \pi x) + ie^{\cos \pi x} \sin(\sin \pi x)$$

Finally we wish to study an extraordinary part of the above function in space.

$$e^{i^{-x+y+z}} + e^{i^{x+y-z}} + e^{i^{x-y+z}} + e^{i^{-x-y-z}} = \text{const} \quad (38)$$

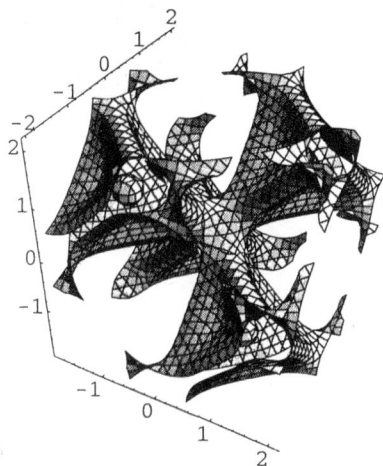

Fig 31. Equation (38) in text gives the complete structure of Zinc Blende, with zinc atoms as spheres in the centre of the tetrahedra.

For a constant of 6 and adding the Re and Im parts a structure as shown in fig 31 is obtained. The group of four tetrahedra sharing one corner is a part of the zinc blende structure. Its composition is ZnS and in the figure the sulphur atoms constitute the corners of the tetrahedra. The zinc atoms are in the centres of tetrahedra, just visible in the figure but really shown to be so in fig 32 (The spheres really belong to the function!).

By changing constant to 6.3 we change curvature and in this case the function opens up catenoids between tetrahedral flatpoints and the spheres. This is seen in figs 33 and 34. The last figure is calculated with -1 as base instead of i.

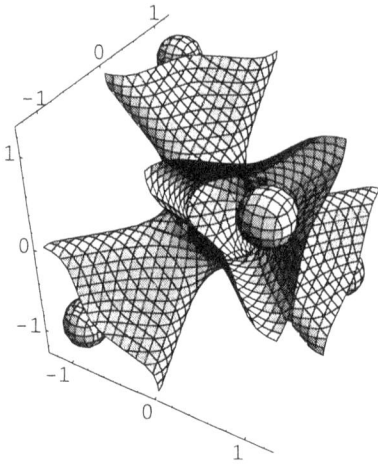

Fig 32. Detail of fig 31.

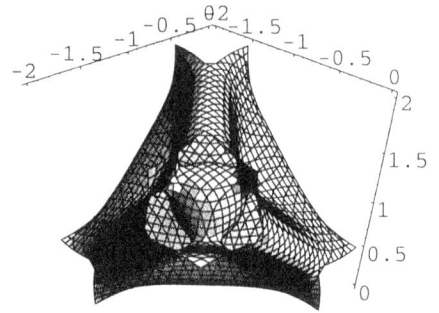

Fig 33. Change of constant gives catenoidic contact between the zinc spheres and the tetrahedra.

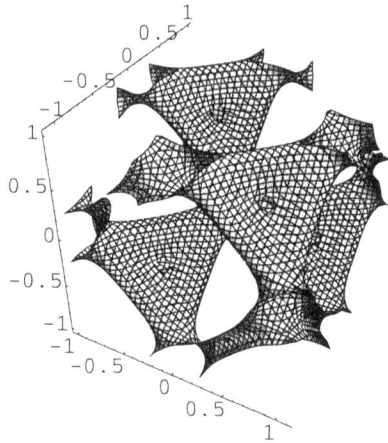

Fig 34. As in fig 33, but bigger part of the structure.

CHAPTER 6

MORE STRUCTURES

The functions we now will study are simple extensions of those studied in the previous chapter:

$$e^{i^{2x}}e^{i^{-2x}} + e^{-i^{2x}}e^{-i^{-2x}} \tag{1}$$

and

$$e^{i^{.2x}} + e^{-i^{.2x}} \tag{2}$$

Equation (1) is identical with

$$e^{\cos x} + e^{-\cos x}$$

which we start with.

$B^{\cos x} + B^{-\cos x}$

The first function to study is

$$e^{\cos 2\pi x} + e^{\cos 2\pi y} + e^{\cos 2\pi z} +$$
$$e^{-\cos 2\pi x} + e^{-\cos 2\pi y} + e^{-\cos 2\pi z} = \text{const} \tag{3}$$

Again we have the *pc* structure for a constant of 6.8 in fig 1. For 7.085 atoms touch each other in a ReO$_3$ similar structure (fig 2) which via catenoids (fig 3, const 7.1) form a nodal P-surface (fig 4, const 7.56) that becomes another ReO$_3$ structure for a constant of 8.17 (fig 5).

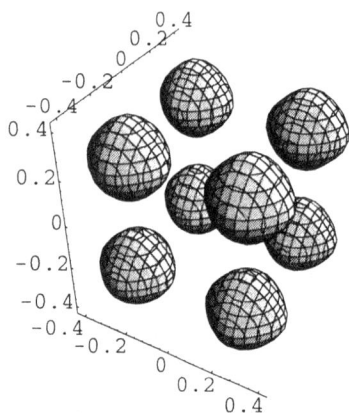

Fig 1. From eq. (3) in text. Const = 6.8 gives a primitive cubic structure.

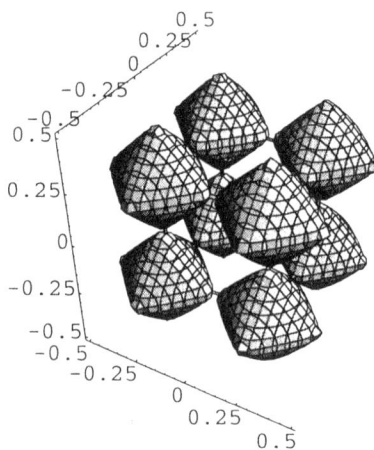

Fig 2. Atoms touch each other for a const of 7.085, which also is a ReO_3 structure.

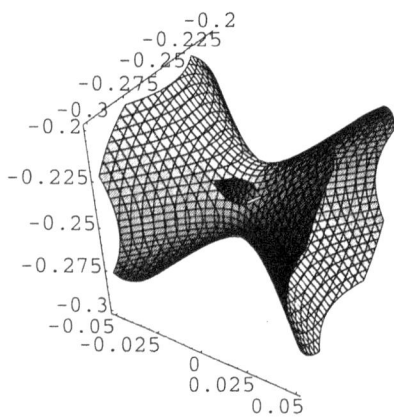

Fig 3. Catenoids develop for a const of 7.1.

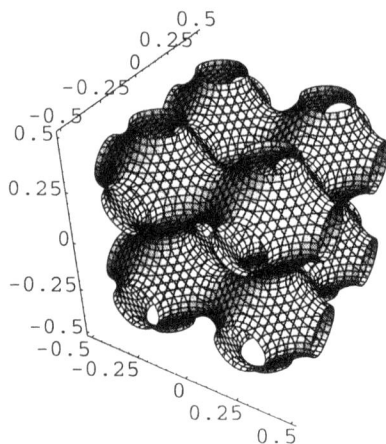

Fig 4. A nodal P- surface is formed for a const of 7.56.

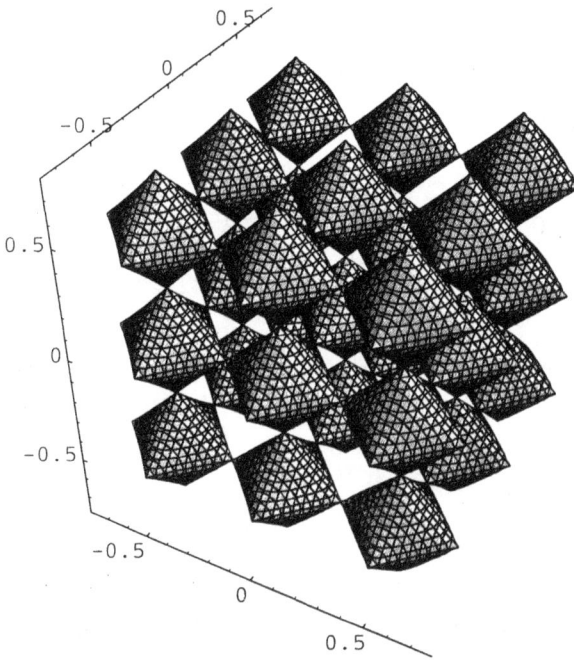

Fig 5. For a const of 8.17 a new ReO₃ structure is formed.

If these two ReO3:s could coexist they would interpenetrate with the nodal surface separating the two. The two ReO3:s are different demonstrating that the nodal surface is splitting space in two non - identical parts.

$$B\,cos(x+y) \;+\; B\text{-}cos(x+y)$$

This gives us the function below which really is a modulation of the rhombic dodecahedron.

$$10^{\cos 2\pi(x+y)} + 10^{-\cos 2\pi(x+y)} +$$

$$10^{\cos 2\pi(x-y)} + 10^{-\cos 2\pi(x-y)} +$$

$$10^{\cos 2\pi(y+z)} + 10^{-\cos 2\pi(y+z)} +$$

$$10^{\cos 2\pi(y-z)} + 10^{-\cos 2\pi(y-z)} + \tag{4}$$

$$10^{\cos 2\pi(x-z)} + 10^{-\cos 2\pi(x-z)} +$$

$$10^{\cos 2\pi(x+z)} + 10^{-\cos 2\pi(x+z)} = \text{const}$$

We now start from a larger constant, 42, and obtain the desired *bcc* packing as shown in fig 6.

These cubes join vertex to vertex with decreasing constant, via catenoids, as shown in fig 7 (constant = 35), and this is the structure of $NiHg_4$.

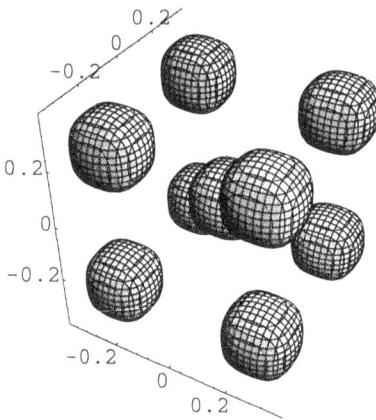

Fig 6. From eq. (4) in text. With a const of 42.0 a cubic body centred arrangements of rounded cubes is obtained.

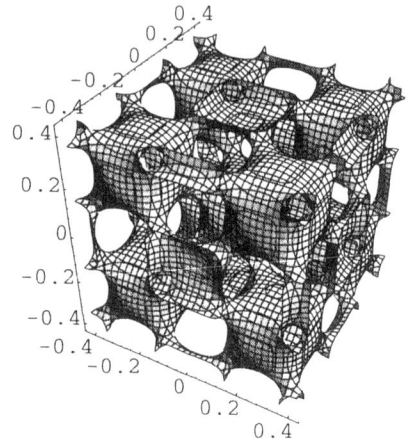

Fig 7. With a const of 35.0 cube corners join via catenoids.

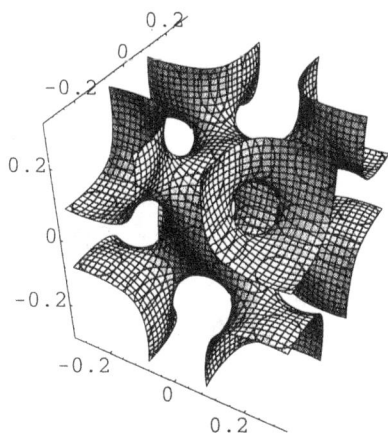

Fig 8. With a const of 33.0 the IWP
surface is demonstrated.

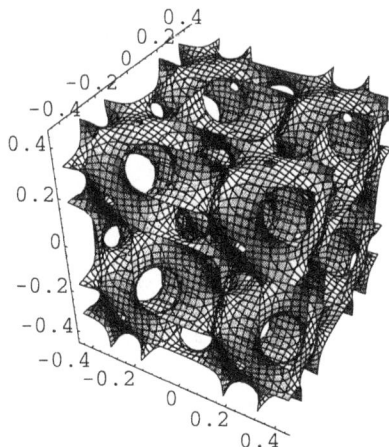

Fig 9. The const is 31.0.

At a constant of 33 we see the beautiful cube with its catenoids
in fig 8 and at 31 we still have the same structure but the
expansion of the catenoids set our perception into action (fig 9).
It is clear why, when the constant changes to 30.7, 28, and 26.8
in figures 10, 11, and 12. The last figure gives the structure of
sodalite again but somewhat better than in last chapter. The
relationship between this structure and that of $NiHg_4$ was most
unexpected.

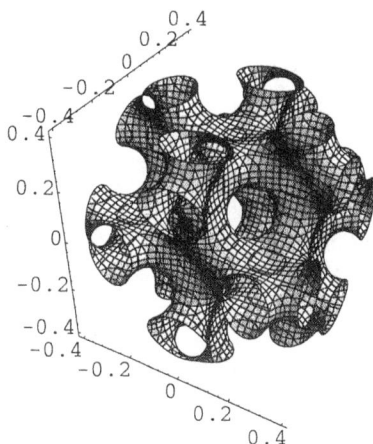

Fig 10. The const is 30.7.

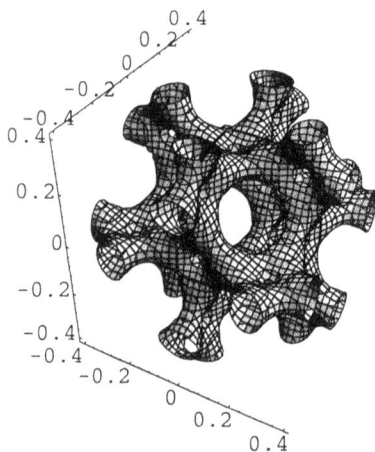

Fig 11. The const is 28.0.

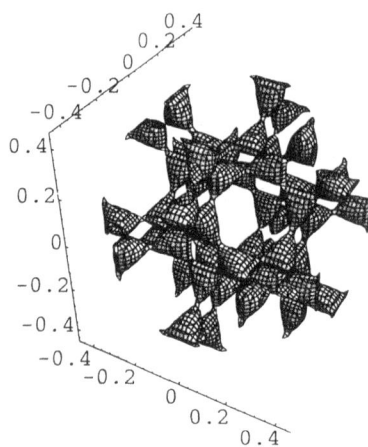

Fig 12. The const is 26.8 and the preceding pictures show the topological transformation to this structure of Sodalite type.

$$B\,cos(x+y+z) \;+\; B\text{-}cos(x+y+z)$$

We start with

$$
\begin{aligned}
&10^{\cos 2\pi(-x+y+z)} + 10^{\cos 2\pi(x+y-z)} + \\
&10^{\cos 2\pi(x-y+z)} + 10^{\cos 2\pi(-x-y-z)} + \\
&10^{-\cos 2\pi(-x-y-z)} + 10^{-\cos 2\pi(x+y-z)} + \\
&10^{-\cos 2\pi(y+z-x)} + 10^{-\cos 2\pi(x-y+z)} = \text{const}
\end{aligned}
\tag{5}
$$

and a constant of 16 gives the Kepler star of cubic close packed spheres as shown in fig 13.

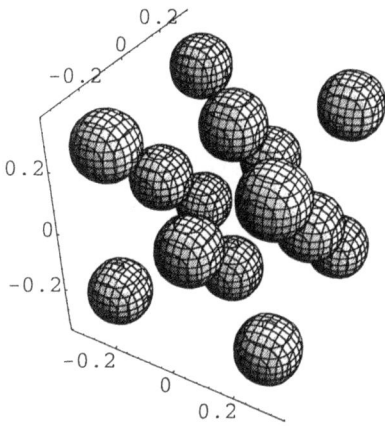

Fig 13. From eq. (5) in text. At a const. of 16.0 the spheres are in a Kepler star of cubic close packing.

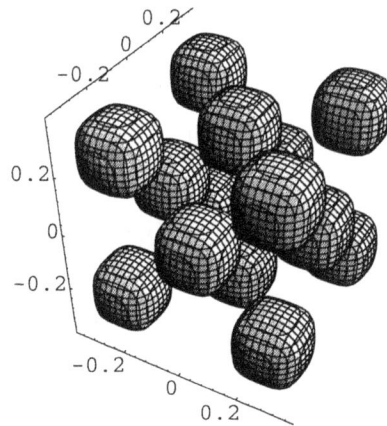

Fig 14. At a const of 20.0 atoms are of cubic shape.

At a constant of 20 the atoms have come closer and have also approached a cubic shape (fig 14).

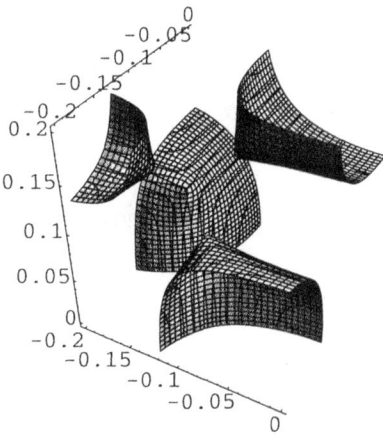

Fig 15. The cubes approach each other in tetrahedral manner at a const of 21.433.

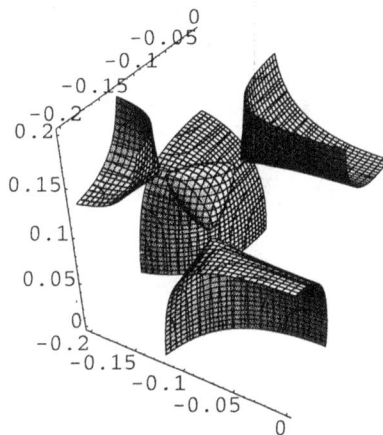

Fig 16. At a const of 21.435 a tetrahedron is formed between the cubes.

The cubes approach each other in a tetrahedral manner as shown in fig 15, at a constant of 21.433, and at 21,435 suddenly a tetrahedron is formed between the cubes (fig 16).

At 21.44 we see from a bigger part of the function that the structure of CaF_2 or TiH_2 is really described (fig 17).

At 21.5 the tetrahedra start to join the cubes and grow a continuos surface as seen in fig 18 and 19.

At a constant of 23.0 the surface formed is a beautiful demonstration of how perpendicular planes can go through each other without intersections (fig 20). This surface is related to the so-called FRD surface. With increasing constant the surface is split up into polyhedra, this time cube-octahedra, which is shown in form of a new structure in fig 21.

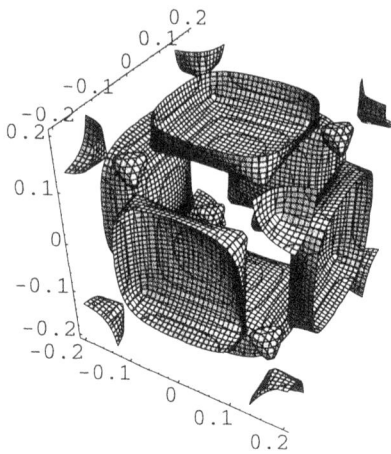

Fig 17. The CaF_2 structure is formed at a const of 21.44.

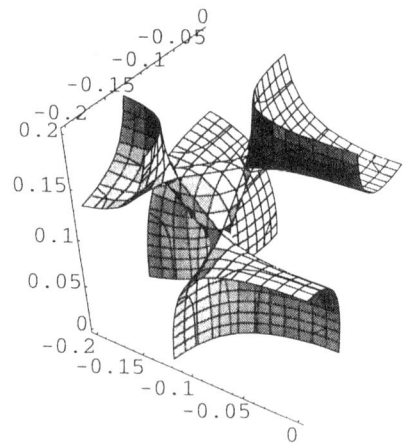

Fig 18. The tetrahedra join to the cubes.

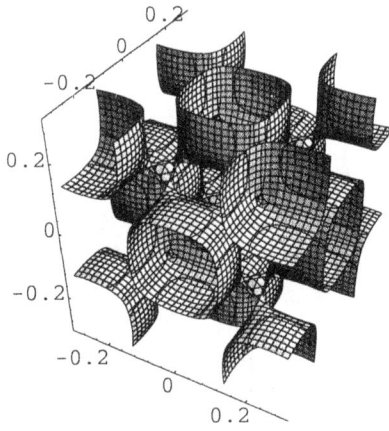

Fig 19. The surface, of FRD, type formed at a const of 21.5.

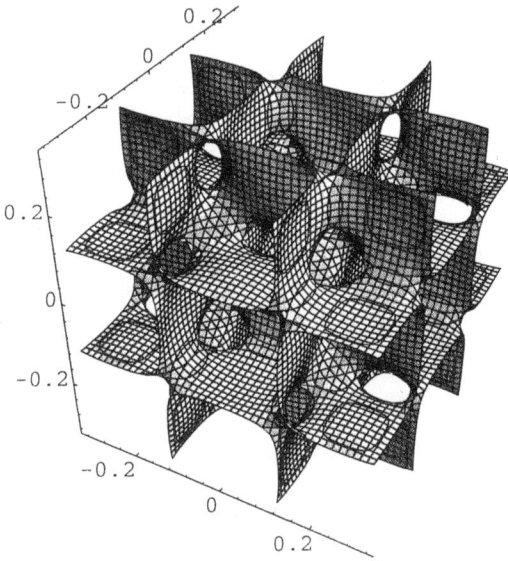

Fig 20. At a const. of 23.0 it is shown how this surface demonstrates that perpendicular panes can go through each other without intersections.

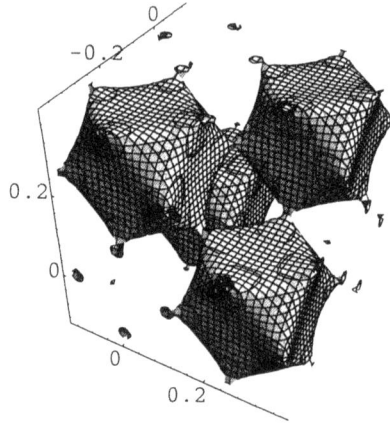

Fig 21. Increasing constant give a
structure of corner sharing cube octahedra.

Again, we repeat, this is similar to what we have seen in chapter
4 but we have performed the study in more detail to show the
differences. For example, the surface in fig 20 is clearly
perpendicular planes, while the one in fig 28 chap 4 is more
rodlike.

Of course we realise that this mathematics is very rich. A
splendid example of this is fig 22 where we have pulled out the
FRD from fig 20, or made it tetragonal, with equation (6)
below. The pleasure of analysis we give to the reader.

$$10^{\cos 2\pi(-x+y+.25z)} + 10^{\cos 2\pi(x+y-.25z)} +$$
$$10^{\cos 2\pi(x-y+.25z)} + 10^{\cos 2\pi(-x-y-.25z)} +$$
$$10^{-\cos 2\pi(-x-y-.25z)} + 10^{-\cos 2\pi(x+y-.25z)} +$$
$$10^{-\cos 2\pi(y+z-.25x)} + 10^{-\cos 2\pi(x-y+.25z)} = 23.2$$

(6)

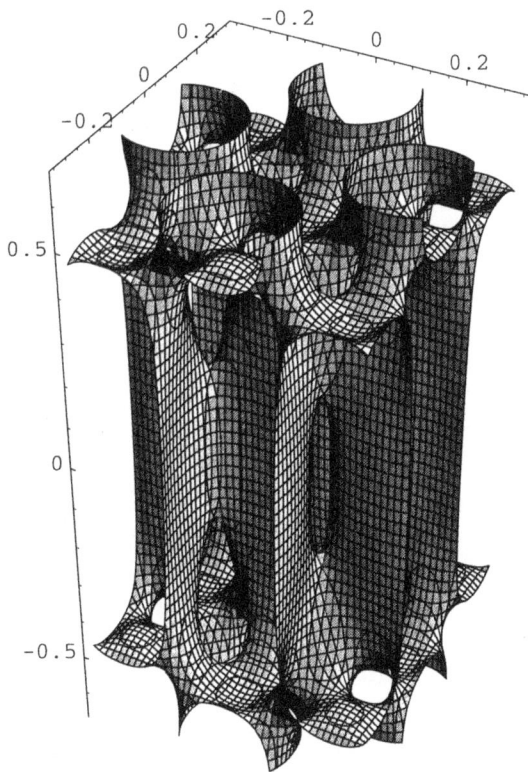

Fig 22. The FRD surface is made tetragonal with eq. (6).

$$B^{i^{2x}} + B^{-i^{2x}}$$

This function is similar to its simpler version earlier in this chapter, but we wish to show some beautiful structures. The first equation is

$$100^{i^{2x}} + 100^{-i^{2x}} + 100^{i^{2y}} + 100^{-i^{2y}} +$$
$$100^{i^{2z}} + 100^{-i^{2z}} = \text{const} \tag{7}$$

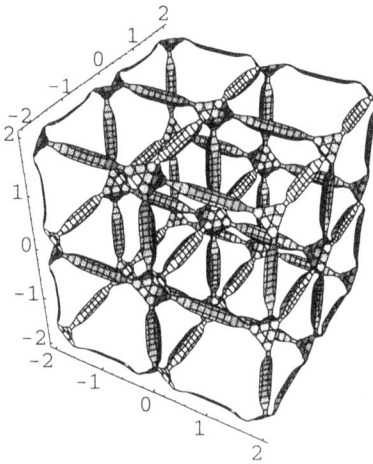

Fig 23. The ReO$_3$ net after eq. (7).

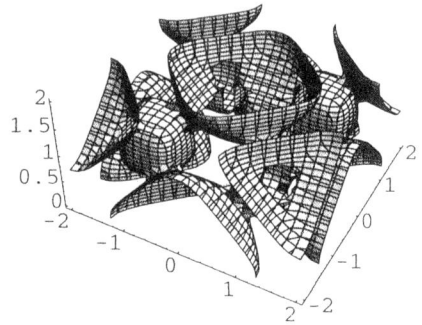

Fig 24. The CaF$_2$ structure after eq. (8).

and its real part gives the remarkable ReO$_3$ net of fig 23 at a constant of 150. Next equation is

$$e^{i^{-x+y+z}} + e^{i^{x+y-z}} + e^{i^{x-y+z}} + e^{i^{-x-y-z}} +$$
$$e^{i^{x-y-z}} + e^{i^{-x-y+z}} + e^{i^{-x+y-z}} + e^{i^{x+y+z}} = \text{const} \tag{8}$$

and for a constant of 8 and its real part we get a structure we seen earlier - the CaF$_2$ structure - but with somewhat larger fluorines. We will now use the rule of addition to cut off a piece of a periodic structure to make an ion or a molecule. We construct the equation

$$e^{i^{-x+y+z}} + e^{i^{x+y-z}} + e^{i^{x-y+z}} + e^{i^{-x-y-z}} + e^{i^{x-y-z}} +$$

$$e^{i^{-x-y+z}} + e^{i^{-x+y-z}} + e^{i^{x+y+z}} + e^{x^2+y^2+z^2} = const \qquad (9)$$

and here the extra term is the sphere. With a constant of 12.2 and the real part of the equation we get fig 25.

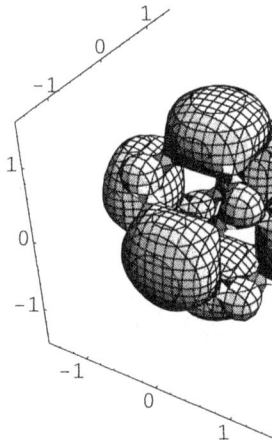

Fig 25. The Mo_6Cl_8 cluster after eq. (9).

We have no longer a periodic structure, the sphere has cut off a unit which is the structure of the famous cluster or ion - Mo_6Cl_8 or Mo_6S_8.

We now want to show the possibilities that the combinations of functions offer. We shall use the functions (10) and (11).

Function (10) is shown in fig 26 where two structures interpenetrate in a simple way. Taken as one structure, it is *fcc* or cubic close packing, with the cubes in a separate *pc* packing. The chemical correspondence is perovskite, $CaTiO_3$, with Ca as cubes and oxygens as the flat bodies. Ti is in the octahedra in perovskite and that is easily traced in fig 26. To conclude, the flat bodies in this structure form two kinds of interstices, the

cube octahedral for the cubes here and for Ca in perovskite and
the octahedral, empty here but with Ti in perovskite.

$$
\begin{aligned}
&10^{\cos 2\pi(-x+y+z)} + 10^{\cos 2\pi(x+y-z)} + \\
&10^{\cos 2\pi(x-y+z)} + 10^{\cos 2\pi(-x-y-z)} + \\
&10^{-\cos 2\pi(-x+y+z)} + 10^{-\cos 2\pi(x+y-z)} + \\
&10^{-\cos 2\pi(x-y+z)} + 10^{-\cos 2\pi(-x-y-z)} + \\
&10^{\cos 2\pi x} + 10^{\cos 2\pi y} + 10^{\cos 2\pi z} + \\
&10^{-\cos 2\pi x} + 10^{-\cos 2\pi y} + 10^{-\cos 2\pi z} = 36.5
\end{aligned}
\tag{10}
$$

$$
\begin{aligned}
&10^{\cos 2\pi(-x+y+z)} + 10^{\cos 2\pi(x+y-z)} + \\
&10^{\cos 2\pi(x-y+z)} + 10^{\cos 2\pi(-x-y-z)} + \\
&10^{-\cos 2\pi(-x+y+z)} + 10^{-\cos 2\pi(x+y-z)} + \\
&10^{-\cos 2\pi(x-y+z)} + 10^{-\cos 2\pi(-x-y-z)} + \\
&10^{\cos 2\pi(x+y)} + 10^{-\cos 2\pi(x+y)} + 10^{\cos 2\pi(x-y)} + \\
&10^{-\cos 2\pi(x-y)} + 10^{\cos 2\pi(y+z)} + 10^{-\cos 2\pi(y+z)} + \\
&10^{\cos 2\pi(y-z)} + 10^{-\cos 2\pi(y-z)} + \\
&10^{\cos 2\pi(x-z)} + 10^{-\cos 2\pi(x-z)} + 10^{\cos 2\pi(x+z)} + \\
&10^{-\cos 2\pi(x+z)} = 58.5
\end{aligned}
\tag{11}
$$

Short edges of the flat bodies react with the edges of the cubes
and a beautiful surface of the Neovius type is formed, for a
constant of 40, as seen in fig 27.

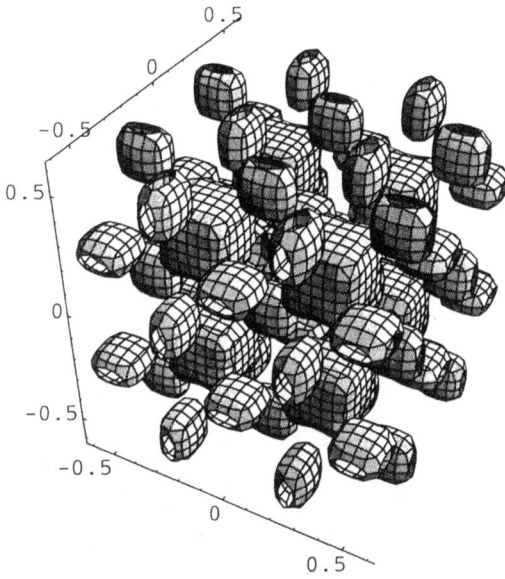

Fig 26. The Perovskite structure after eq. (10) and a const of 36.5.

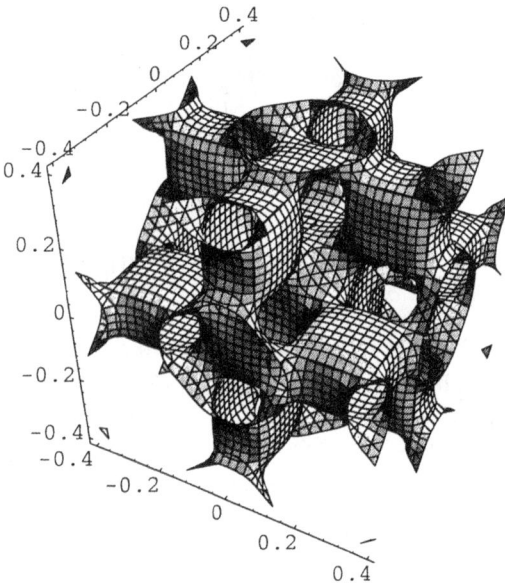

Fig 27. The Neovius type surface at a const of 40.0.

Equation 11 gives the structure in fig 28, which is a primitive
arrangement of cubes interpenetrating a P- surface or ReO_3 net
so we have a new description of the perovskite structure again.
The prehistory is a most remarkable surface shown in fig 29 and
the constant is 56. Here the P - surface and the cubes are fused
together to give a topology that makes the inside of the surface -
the cavities - belong to the outside! The prehistory to this surface
is a structure like the one in fig 26 but without cubes (const 50).

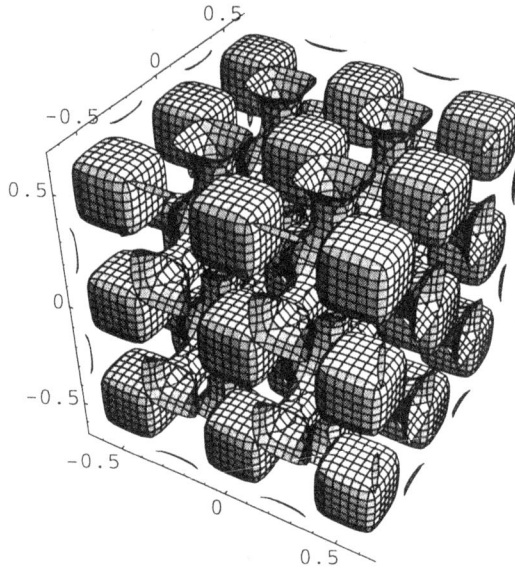

Fig 28. The Perovskite structure after eq. (11), and a const of 58.5.

We cannot resist to show a remarkable part of an other surface.
The combination is like in equation (10) but with $10^{\cos 2\pi 2x}$
instead of $10^{\cos 2\pi x}$. The fundamental unit is shown in fig 30,
an octahedron with 12+6 catenoids. The complete surface is
shown in fig 31, the constant is 36, while it is 36.7 for fig 30.
The surface really describes the zincblende structure, but in a
more complicated way than in chapter 5.

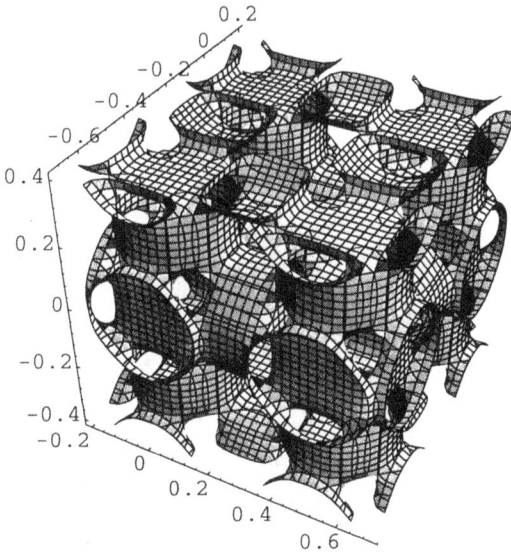

Fig 29. The prehistory of the surface in fig 28, at a const of 56.0.

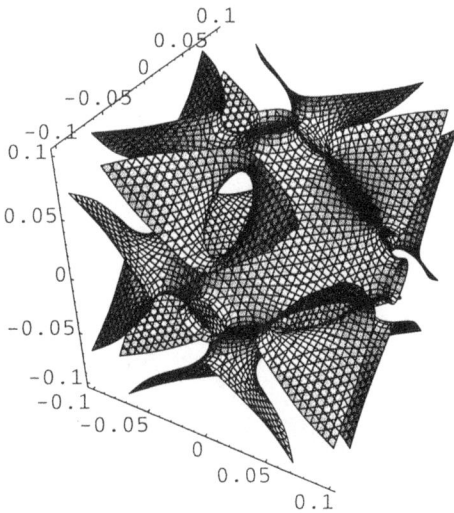

Fig 30. The octahedron with 18 catenoids.

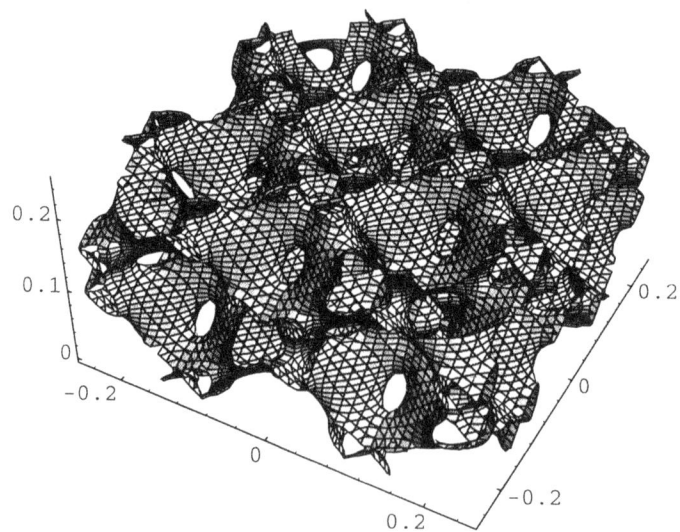

Fig 31. The Zinc Blende structure again.

CHAPTER 7

HEXAGONAL STRUCTURES

We start with the trigonal bipyramid, and using *sine* gives us hexagonal close packing, **hcp.**

$$10^{\sin 2\pi(\frac{1}{2}x+\frac{1}{2\sqrt{3}}y+\frac{1}{\sqrt{6}}z)} + 10^{\sin 2\pi(\frac{1}{2}x+\frac{1}{2\sqrt{3}}y-\frac{1}{\sqrt{6}}z)} +$$

$$10^{\sin 2\pi(-\frac{1}{2}x+\frac{1}{2\sqrt{3}}y+\frac{1}{\sqrt{6}}z)} + 10^{\sin 2\pi(-\frac{1}{2}x+\frac{1}{2\sqrt{3}}y-\frac{1}{\sqrt{6}}z)} + \qquad (1)$$

$$10^{\sin 2\pi(-\frac{1}{\sqrt{3}}y+\frac{1}{\sqrt{6}}z)} + 10^{\sin 2\pi(-\frac{1}{\sqrt{6}}z-\frac{1}{\sqrt{3}}y)} = \text{const}$$

Short hand notation gives

$$10^{\sin(\text{trig\_bipyr})} = \text{const}$$

with face vectors

$$(\pm\frac{1}{2},\frac{1}{2\sqrt{3}},\pm\frac{1}{\sqrt{6}}),\ (0,-\frac{1}{\sqrt{3}},\pm\frac{1}{\sqrt{6}}).$$

Many chemical structures are related to this structure type, and it has become convenient to use another face vector set containing some of the deviations, that so commonly occurs, from the ideal c/a:

$$(\frac{1}{2},\frac{1}{2\sqrt{3}},0.41) \qquad (2)$$

This is accordingly used below.

Fig 1 shows us, for a constant of 6, an octahedron of bodies that belong to the hexagonal close packed structure. A good rule of thumb says that octahedra in this structure share faces along c and edges in the ab plane. In the **ccp** structure they share edges everywhere.

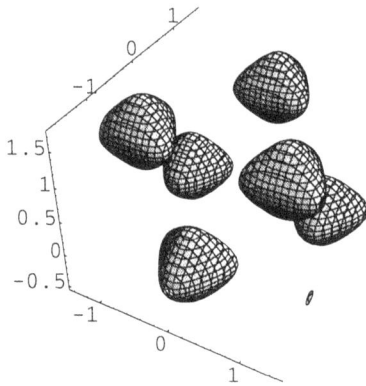

Fig 1. Hexagonally close packed array of bodies, after eq. (1), and a const of 6.0.

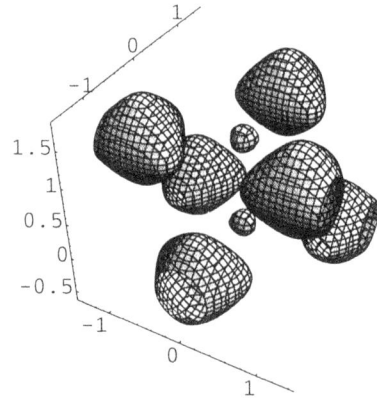

Fig 2. Smaller extra bodies appear at a const of 9.0.

Increasing the constant to 9 gives the structure shown in fig 2, which has strong similarities to that of BCl_3. If every second of the smaller atoms were deleted, regularly along c, it would be identical.

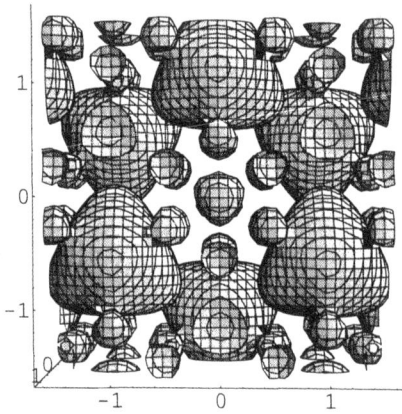

Fig 3. A structure very close to that of $Y(OH)_3$ appear at a const of 12.0.

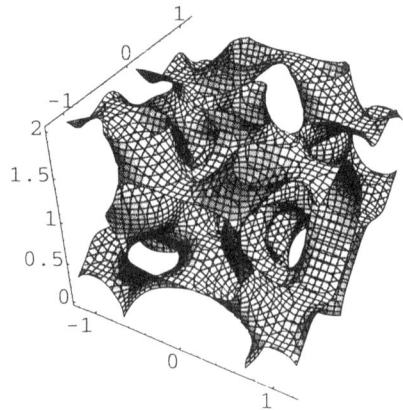

Fig 4. A net develops at a const of 19.0.

Increasing the constant to 12 gives a structure that is very close
to that of Y(OH)$_3$, as shown in fig 3. The larger trigonal
bipyramidal bodies are Y atoms, and the smaller are oxygens.
The 'borons' are still there and now in the centres of the oxygen
octahedra.

Augmenting the constant means the development of some
complicated surfaces, we show one in fig 4 for a constant of 19
where a net character is starting to develop. This is more
developed for a constant of 20 and we show now with a few
figures a very remarkable surface. In order to understand it we
need to go deep into some structural chemistry.

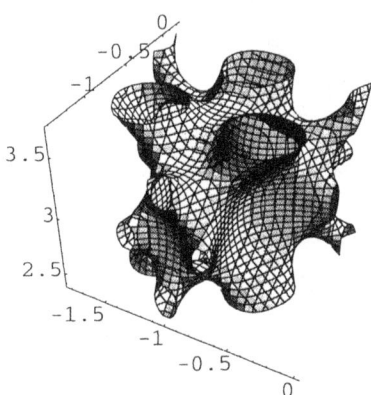

Fig 5. 12+6 catenoids of a hexagonally
close packed atom.

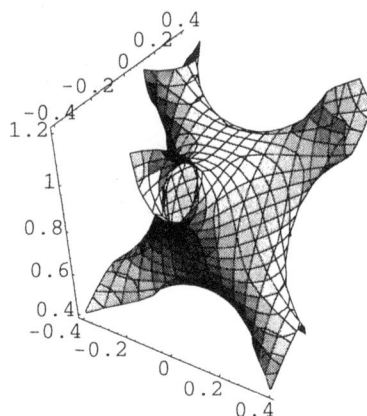

Fig 6. Six catenoids around an
interstitial oxygen in *hcp*.

Close packed metals have so called octahedral interstices, as in
fig 1, and several of them dissolve oxygen, carbon, or nitrogen
into that space. This very important property constitutes a great
part of the metallurgy. The metal titanium dissolves oxygen up
to TiO$_{0.5}$ and the oxygens occupy half of these octahedral sites,
in random at higher temperature. In titanium each metal atom has
12 neighbours with the geometry of the hexagonal cube
octahedron described in chapter 1. With oxygen dissolved into it
there are six more neighbours for each titanium. The geometry
for all these 12+6 neighbours is provided by this surface of

hexagonal close packing. In fig 5 we show the catenoidic surrounding of a metal atom. There are six heavy catenoids, three up, three down that join to other metal atoms. There are six more, thinner, that are in the **ab** plane and also join to other metal atoms. Finally there are six more, also thin, three up, three down, that demonstrate the bonding to the oxygens. The geometrical surrounding of an oxygen is shown in fig 6.

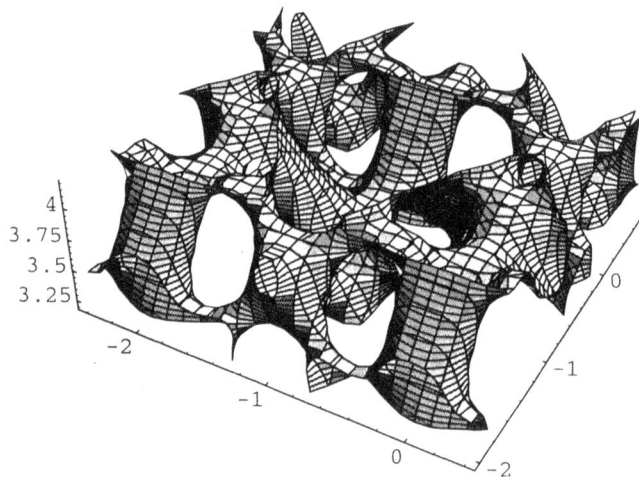

Fig 7a. The structure of NiAs showing *hcp* As atoms and interstitial Ni.

This can also be compared with the NiAs structure, which has arsenic atoms in **hcp** and nickel atoms in octahedral interstices. In fig 7a we see in a part of the structure how the octahedra (oxygens) join to the hexagonal cube octahedra (titanium). In fig 7b we see the projection along **c**.

We will also show this function for a value of the constant of 23 as the net character is more pronounced. This is shown in fig 8 and several of the catenoids have been closed. The NiAs description is clear now with the small balls being nickel atoms.

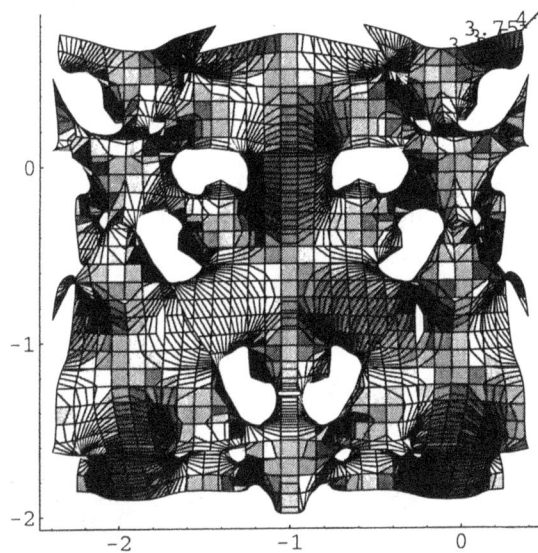

Fig 7b. The same structure projected along **c**.

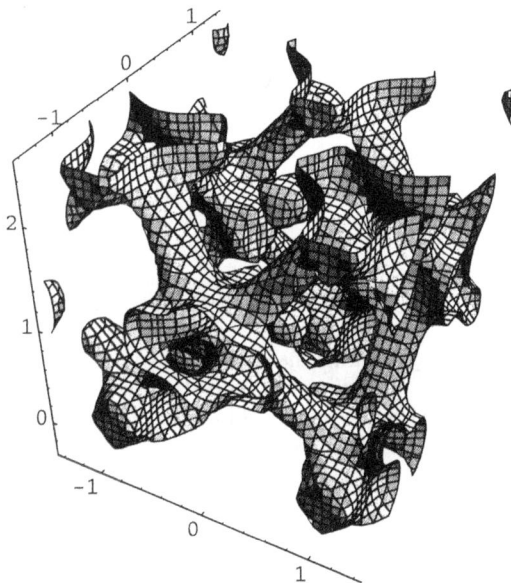

Fig 8. At a const of 23.0 the net character is more developed, catenoids are closed and the Ni atoms appear as small spheres.

With equation (3) below we will show some really beautiful structures. Face vectors as in (2).

$$10^{\cos(\text{trig}\_\text{bipyr})} + 10^{-\cos(\text{trig}\_\text{bipyr})} = \text{const} \qquad (3)$$

A study of this function follows the usual pattern and we start at a constant of 23 which gives the AlB$_2$ structure, as shown in fig 9.

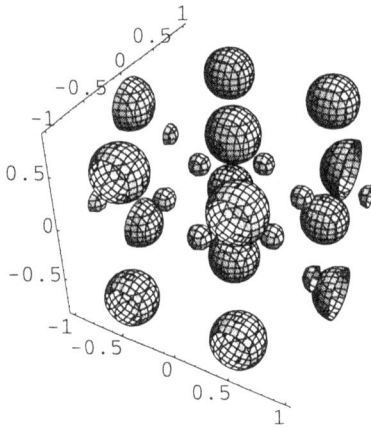

Fig 9. Eq. (3) gives the AlB$_2$ structure with a const of 23.0.

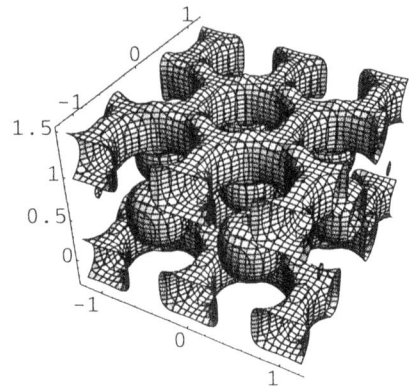

Fig 10. At a const of 31.0 the boron atoms have condensed to graphite like layers.

The small atoms are borons and at a constant of 31 the boron atoms have fused together to graphite like layers as shown in fig 10. Augmenting the constant make the spheres (Al atoms in AlB$_2$) react with the layers to form a most splendid surface at 34.6, shown in fig 11a and b, and that could be named hexagonal FRD.

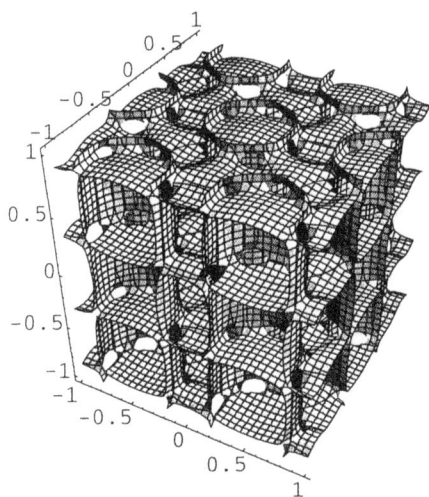

Fig 11a. At 34.6 the larger Al atoms have
fused with the layers to form a hexagonal
variant of the FRD.

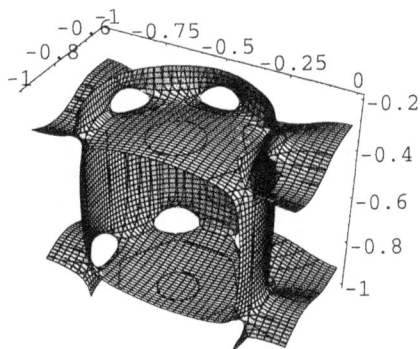

Fig 11b. Detail of the surface.

By scaling the z-partitioning in the function above, from 0.41 to
resp. 0.1 and 1.5 (thus elongating and compressing the trigonal
bipyramids) we can demonstrate the *rod* and the *plane* character
of this surface. In fig 12 we see how planes go through rods
without intersection and in fig 13 and fig 14 we see how rods go
through planes without intersections.

To get some more classic hexagonal structures we need to go to
the trigonal prism with the following equation:

$$10^{\cos 2\pi(x+\frac{\sqrt{3}}{3}y)} + 10^{\cos 2\pi\frac{2\sqrt{3}}{3}y} + 10^{\cos 2\pi(\frac{\sqrt{3}}{3}y-x)} + \qquad (4)$$

$$10^{-\cos 2\pi z} + 10^{\cos 2\pi z} = \text{const}$$

For a constant of 8 we get the honeycomb net in figures 15 and
16.

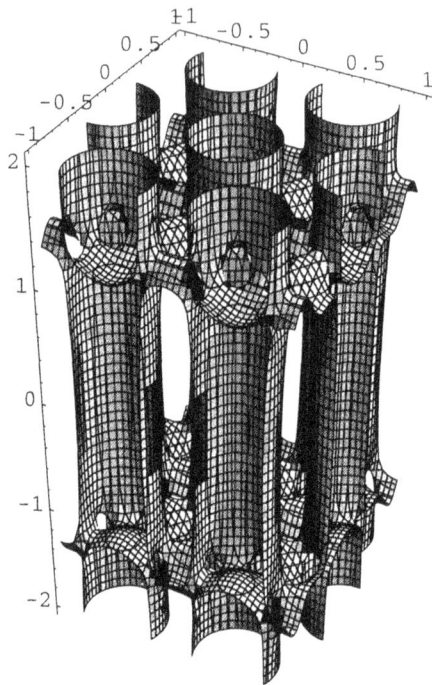

Fig 12. By changing *c/a* is shown how rods go through planes without intersecting.

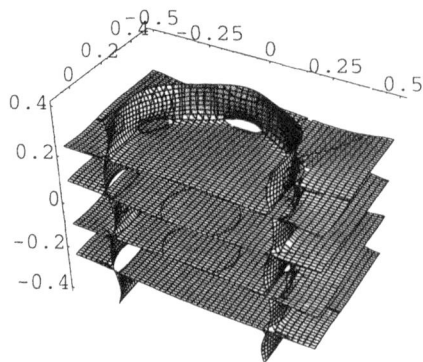

Fig 13. Planes go through rods without intersections.

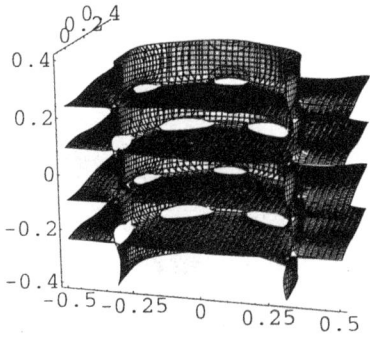

Fig 14. Detail of fig 13.

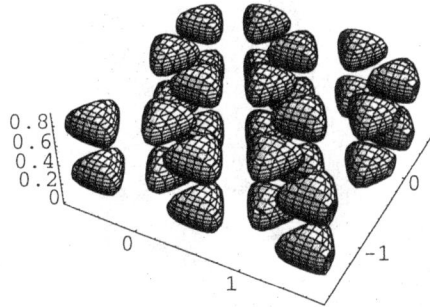

Fig 15. Eq. (4) gives a honeycomb net.

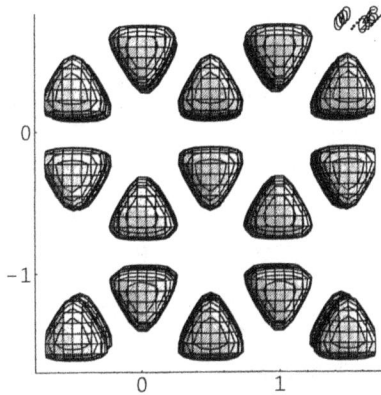

Fig 16. Projection along c of structure in fig. 15.

Increasing the constant to 18 shows a surface emerging from the fusion of trigonal prismoids, which is shown in figures 17 and 18, and is of course the hexagonal correspondent to the cubic P.

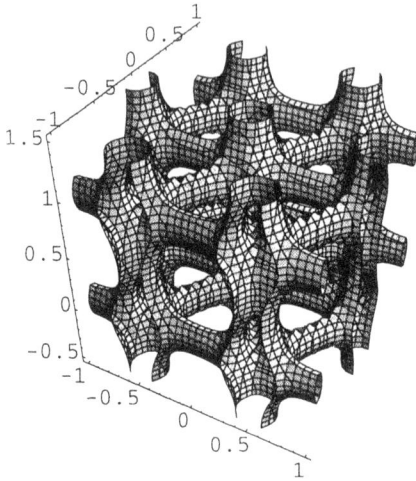

Fig 17. Increasing constant gives a surface that is the hexagonal correspondent to cubic P.

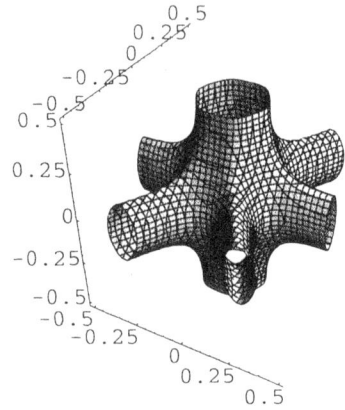

Fig 18. Detail of fig 17.

Equation (5) gives finally some nice structures.

$$10^{\cos 2\pi(x+\frac{\sqrt{3}}{3}y)} + 10^{\cos 2\pi \frac{2\sqrt{3}}{3}y} + 10^{\cos 2\pi(\frac{\sqrt{3}}{3}y-x)} +$$
$$10^{-\cos 2\pi z} + 10^{\cos 2\pi z} + 10^{-\cos 2\pi(x+\frac{\sqrt{3}}{3}y)} +$$
$$10^{-\cos 2\pi \frac{2\sqrt{3}}{3}y} + 10^{-\cos 2\pi(\frac{\sqrt{3}}{3}y-x)} + 10^{-\cos 2\pi z} +$$
$$10^{\cos 2\pi z} = \text{const}$$

(5)

A constant of 20 gives a fused honeycomb structure in form of separate layers. This is made especially pronounced by changing the terms $10^{\cos 2\pi z}$ to $10^{\cos 3\pi z}$ and with a constant of 22 we get fig 19. The structural correspondence is boron nitride. The boundaries have been chosen to give the split version. Back to equation (4) and a constant of 33 we get fusion between the graphitic layers to the very remarkable structure in fig 20. It does not seem to have a correspondent chemical structure, but is of course topologically related to the net in fig 17.

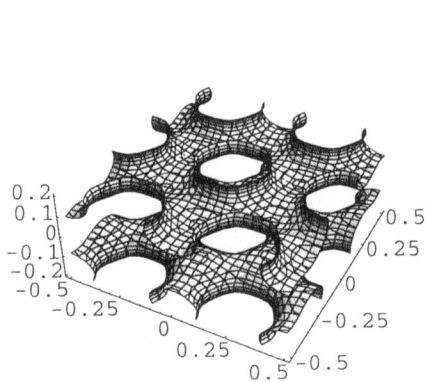

Fig 19. Eq. (5) and a const of 20.0 gives separate honeycomb layers, and the structure is that of BN.

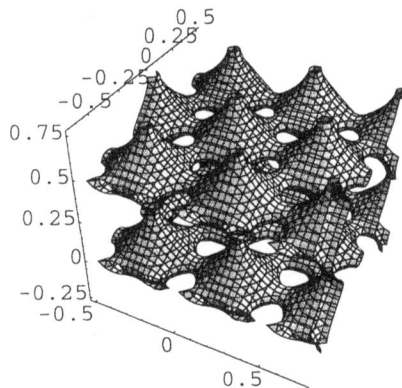

Fig 20. The structure occurring at a const of 33.0.

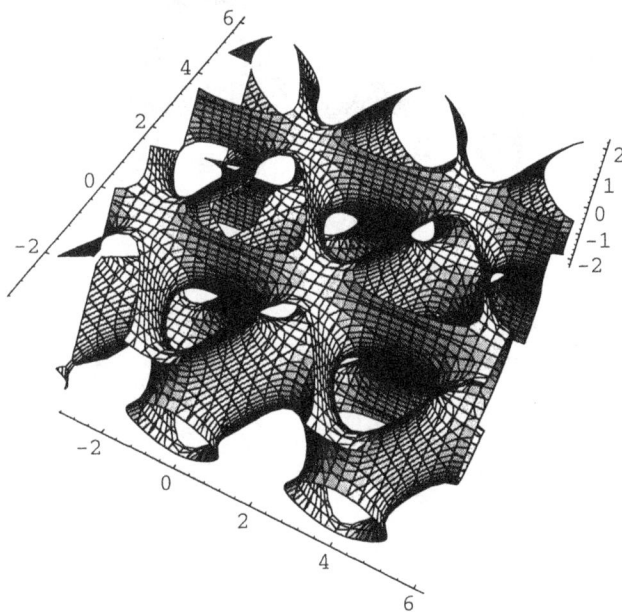

Fig 21a. The Schwartz hexagonal nodal(H) surface after eq. (6).

It was pointed out to us by Sven Lidin that a cosine modulation
of half the hexagonal prism would give the nodal surface of the
exponential surface of fig 17. The general formula is:

$$i^{x+\frac{\sqrt{3}}{3}y} + i^{2\frac{\sqrt{3}}{3}y} + i^{\frac{\sqrt{3}}{3}y-x} + i^z = 0 \qquad (6)$$

Its Real part corresponds then to the surface mentioned while,
surprisingly, the Im part is the H - nodal surface, as shown in
fig 21a and b.

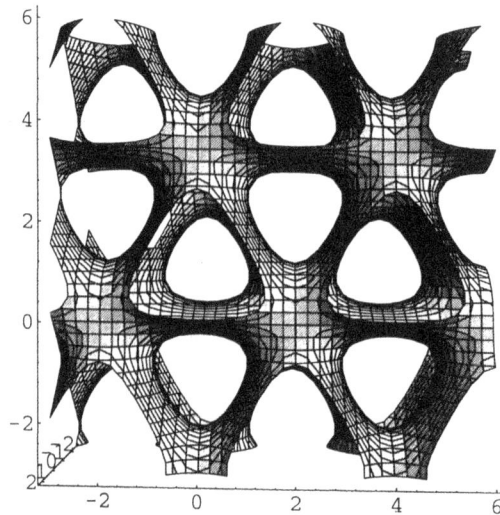

Fig 21b. The H-surface projected after **c**.

CHAPTER 8

CONCENTRIC STRUCTURES

We started with the functions of type that B^x gave polyhedra etc., and i^x that gave the simple nodal surfaces. The B^{i^x} gave structures and ordinary translation - or isometry. We will now study a function that gives the other similarity - the dilatation - the repetition is only near conformal.

$i^x B^x$

We start to study this simple function in two dimensions.

$$i^{2x}e^x + i^{2y}e^y = 0 \tag{1}$$

In figs 1, 2, and 3 we see the Re, Re+Im, and Im parts.

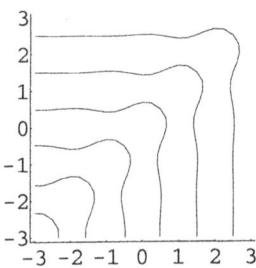

Fig 1. Eq. (1), Re part.

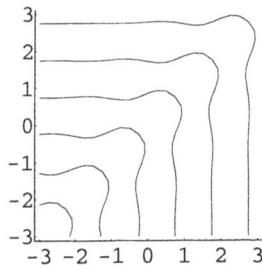

Fig 2. Eq. (1), Re+Im part.

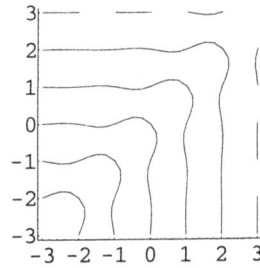

Fig 3. Eq. (1), Im part.

'Square' corners are repeated in a dilative manner. This is more obvious using the equation

$$i^{2x}e^x + i^{2y}e^y + i^{2x}e^{-x} + i^{2y}e^{-y} = 0 \tag{2}$$

and we have plotted the Re, Re+Im, and Im parts in figs 4, 5, and 6.

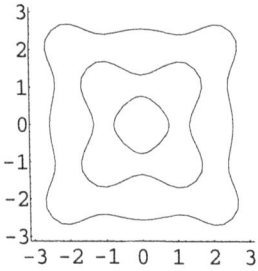

Fig 4. Eq. (2), Re part.   **Fig 5.** Eq. (2), Re+Im part.   **Fig 6.** Eq. (2), Im part.

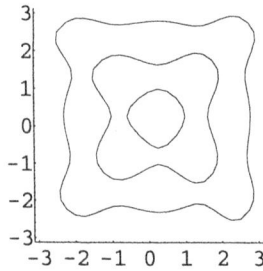

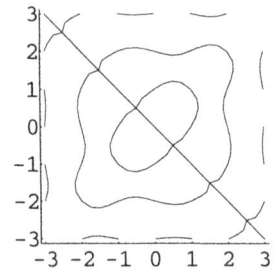

A close look at fig 5 reveals it to be asymmetric as shown in fig 7, and in fig 8 we have just increased the boundaries of fig 4, to show the dilatation.

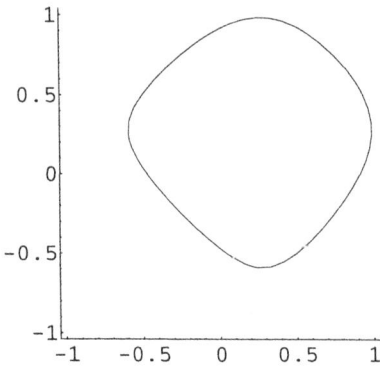

Fig 7. A close look of fig 5.   **Fig 8.** Increased boundaries of fig 4.

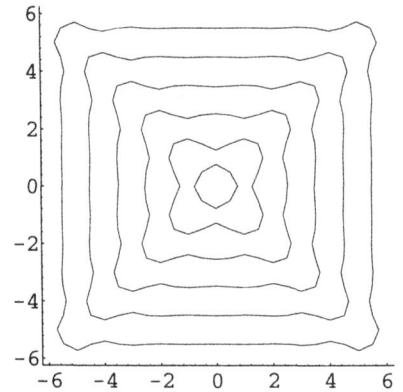

Now we go to three dimensions with the equation

$$i^x e^x + i^y e^y + i^z e^z = 0 \tag{3}$$

Fig 9 shows a 'cube corner' periodically repeated with dilative similarity. This is the Real part of (3), the Im and Im+Re are similar.

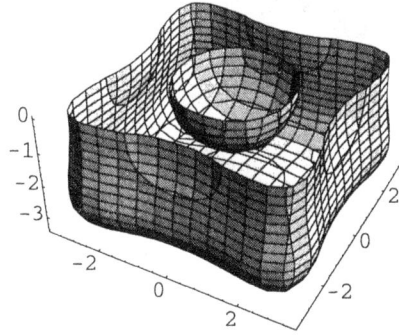

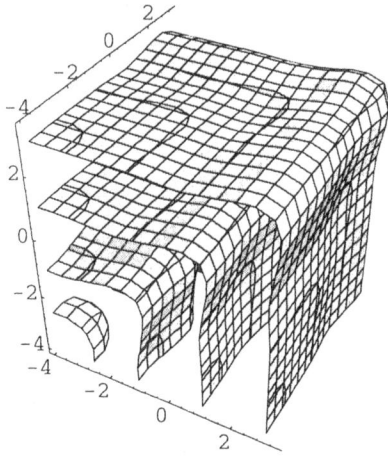

Fig 9. Eq. (3), concentric periodicity of a   **Fig 10.** Eq. (4), Re part and a const. of 0.
cube corner.

Using the complete function

$$i^x e^x + i^y e^y + i^z e^z + i^x e^{-x} + i^y e^{-y} + i^z e^{-z} = 0 \quad (4)$$

we get again the concentric structures which we also seen
before[18]. We will only study the core of this function as that is
where it is interesting. The first picture in fig 10 is for a constant
of 0 and with the boundaries chosen we get a cube with a central
polyhedron. Augmenting constant gives catenoidic contact
between the polyhedra and a beautiful structure reminding about
the electron structure of the molecule $B_6H_6^{2-}$ [21]. This is
shown for the constants 8, 12, and 14 in figs 11, 12 and 13.

The Re + Im of (4) and a constant of 6 gives the surprising
result in fig 14a, b, c. This is obviously a centaur polyhedron of
the solids of fig 11 and 10.

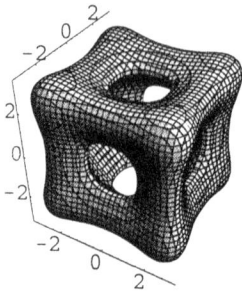

Fig 11. Eq. (4), Re part and a const. of 8.

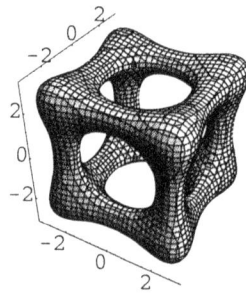

Fig 12. Eq. (4), Re part and a const. of 12.

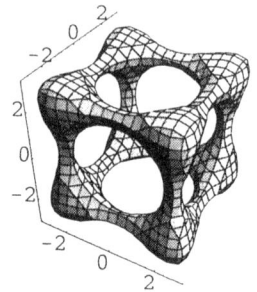

Fig 13. Eq. (4), Re part and a const. of 14.

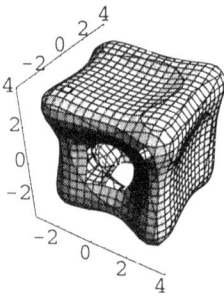

Fig 14a. Eq. (4), const of 6.0, and Re+Im parts gives a centauer polyhedron.

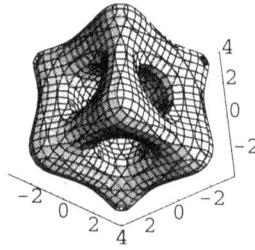

Fig 14b. Different view of a.

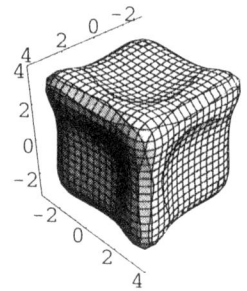

Fig 14c. An other view.

In figs 15 we see the Im part of (4). The function is symmetric for a function with a constant of 0 (a) but not for a constant of 2 (b).

Using the real part of equation (5) below gives fig 16, a tetrahedron continually built around the D surface, and similar in topology to the electronic structure of the molecule B_4H_4 [21].

$$i^{0.5(-x+y+z)} 5^{-x+y+z} + i^{0.5(x+y-z)} 5^{x+y-z} + i^{0.5(x-y+z)} 5^{x-y+z} + i^{0.5(-x-y-z)} 5^{-x-y-z} = 7.1 \qquad (5)$$

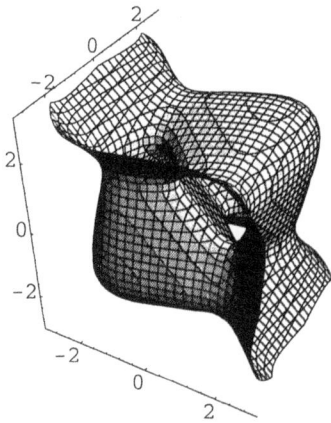

Fig 15a. Eq. (4), const of 0.0 and Im
part gives an opened 'cube corner'.

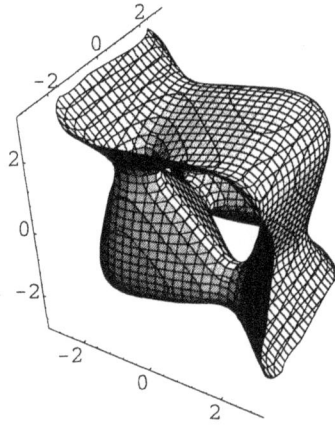

Fig 15b. Same as a, but with
const=2.0.

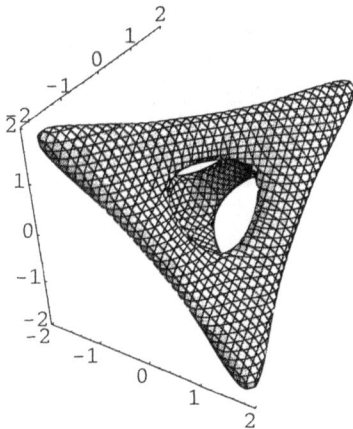

Fig 16. Eq. (5). Re part gives a
tetrahedral structure.

The octahedral character and the real part of equation (6) give us the beautiful structure shown in fig 17 and which we propose as the electronic structure of $B_8H_8^{2-}$.

$$i^{0.5(-x+y+z)}10^{-x+y+z} + i^{0.5(x+y-z)}10^{x+y-z} +$$
$$i^{0.5(-x-y-z)}10^{-x-y-z} + i^{0.5(x+y+z)}10^{x+y+z} + \qquad (6)$$
$$i^{0.5(x-y+z)}10^{x-y+z} + i^{0.5(x-y+z)}10^{-(x-y+z)} = 24$$

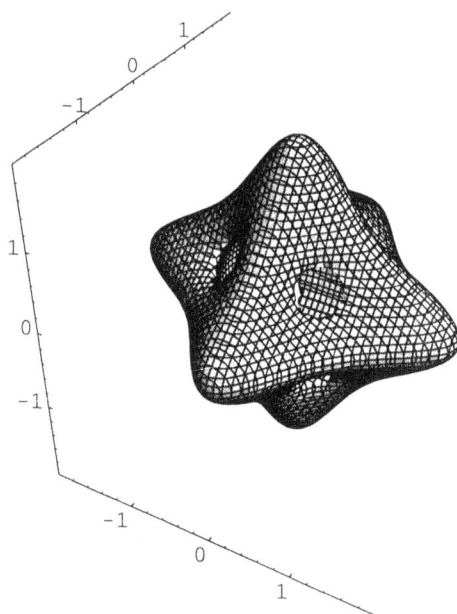

Fig 17. Eq. (6). Re part gives an octahedral structure.

We will also do another Platonic solid - the dodecahedron. Its
equation is:

$$
\begin{aligned}
& i^{0.66(\tau x+y)}10^{\tau x+y} + i^{-0.66(\tau x+y)}10^{\tau x+y} + \\
& i^{0.66(-\tau x+y)}10^{-\tau x+y} + i^{-0.66(-\tau x+y)}10^{-(-\tau x+y)} + \\
& i^{0.66(\tau y+z)}10^{\tau y+z} + i^{-0.66(\tau y+z)}10^{-(\tau y+z)} + \\
& i^{0.66(-\tau y+z)}10^{-\tau y+z} + i^{-0.66(-\tau y+z)}10^{-(-\tau y+z)} + \\
& i^{0.66(-x+\tau z)}10^{-x+\tau z} + i^{-0.66(-x+\tau z)}10^{-(-x+\tau z)} + \\
& i^{0.66(x+\tau z)}10^{x+\tau z} + i^{-0.66(x+\tau z)}10^{-(x+\tau z)} + 9500 = 0
\end{aligned}
\tag{7}
$$

The real part gives the most beautiful surface as seen in fig 18a
and b. The molecule is $B_{12}H_{12}^{2-}$ [21].

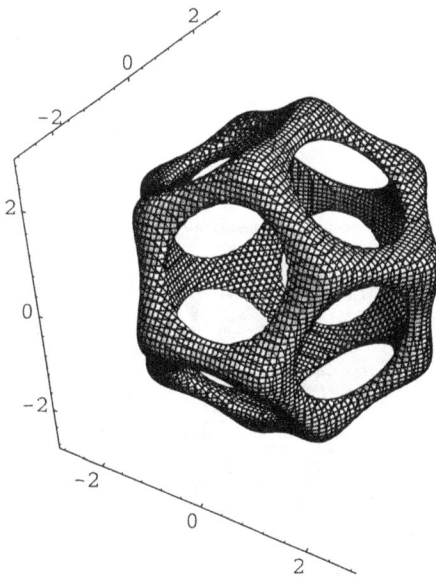

Fig 18a. Eq. (7). Re part gives a dodecahedral structure.

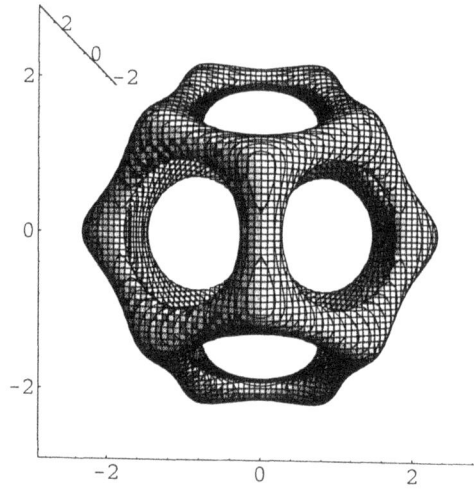

Fig 18b. A projection of a.

For a slightly different constant, 10000, we see parts of the next shell, beautifully demonstrating the cubic symmetry of a Platonic solid, as shown in fig 19.

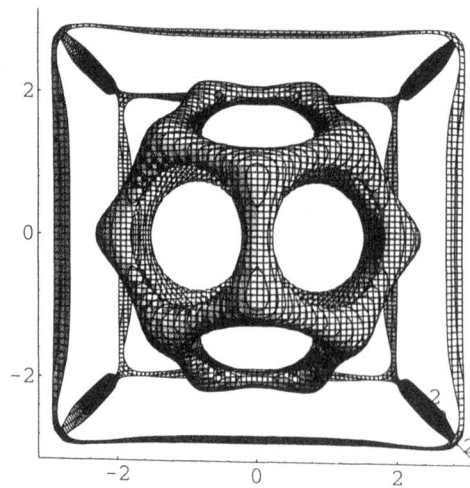

Fig 19. Greater boundaries give next shell.

At last we shall demonstrate this conformal symmetry using the rhombic dodecahedron and the equation is:

$$i^{0.5(x+y)}10^{x+y} + i^{0.5(x+y)}10^{-(x+y)} +$$
$$i^{0.5(-x+y)}10^{(-x+y)} + i^{0.5(-x+y)}10^{-(-x+y)} +$$
$$i^{0.5(y+z)}10^{(y+z)} + i^{.05(y+z)}10^{-(y+z)} +$$
$$i^{0.5(-y+z)}10^{(-y+z)} + i^{0.5(-y+z)}10^{-(-y+z)} +$$
$$i^{0.5(-x+z)}10^{(-x+z)} + i^{0.5(-x+z)}10^{-(-x+z)} +$$
$$i^{0.5(x+z)}10^{(x+z)} + i^{0.5(x+z)}10^{-(x+z)} + 20000 = 0$$

$$(8)$$

The real part is given in fig 20.

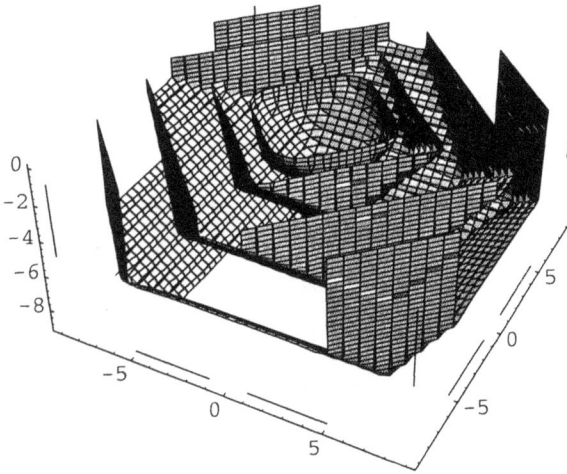

Fig 20. Demonstrating the conformal symmetry using the rhombic dodecahedron and eq. (8).

CHAPTER 9

POTPOURRI

The property of the exponential scale saying that separate functions can be added to become a new continuos function is a fundament of this work. We like to demonstrate or show that it might be useful also outside chemistry or ordinary crystal structure science.

We start with the cone in the first equation:

$$10^{x^2+y^2-z^2} = 1000 \qquad\qquad (1)$$

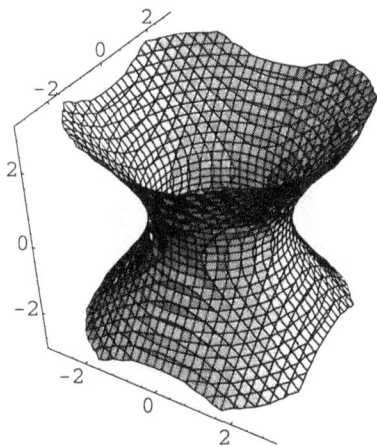

Fig 1. The eq. of a double cone (1) gives a catenoid.

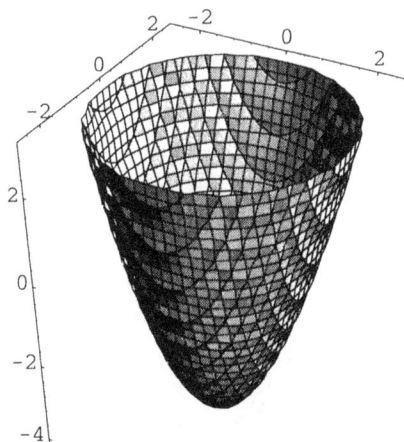

Fig 2. Eq. (2) gives a cone.

Fig 1 gives a beautiful catenoid, a cone 'rounded off'. But we can also see this as catenoidic opening of the sphere - because of z^2 - in two directions. So we use just z:

$$10^{x^2+y^2-z} = 10000 \tag{2}$$

And we get a closed cone or bucket, in fig 2.

Next thing to do is to add a sphere to equation (1)

$$10^{x^2+y^2-z^2} + 10^{0.3(x^2+y^2+z^2)} = 200 \tag{3}$$

And we get a bollard in fig 3. We can open a sphere differently:

$$10^{x^2+y^2+z^2} - 10^{0.8x^3} = 300 \tag{4}$$

This gives a splendid lamp in fig 4.

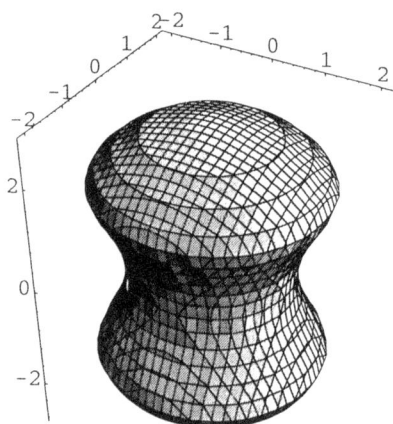

Fig 3. The double cone closed with a sphere.

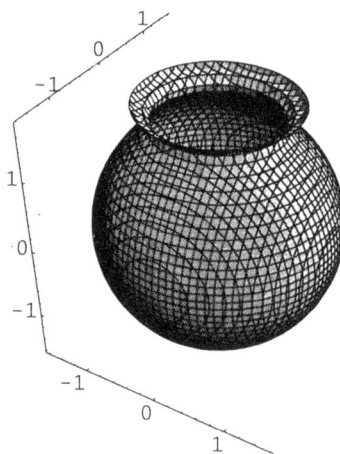

Fig 4. Opening of a sphere gives a lamp.

Now we shall add a helicoid to the cylinder. First the equation of the helicoid

$$z\cos 4x - y\sin 4x = 0 \tag{5}$$

giving fig 5.
And then we add the cylinder and the helicoid

$$10^{0.5(y^2 + z^2)} + 10^{z\cos 4x - y\sin 4x} = 2 \tag{6}$$

which gives the screw in fig 6

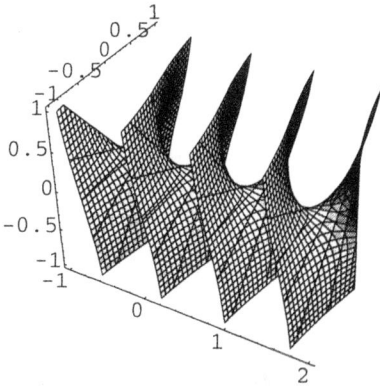

Fig 5. The helicoid.

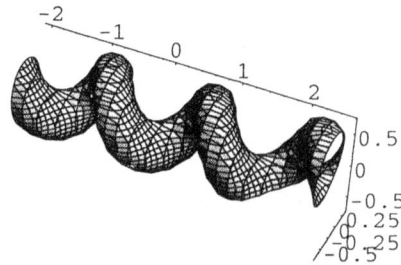

Fig 6. Adding a cylinder to the helicoid gives the screw.

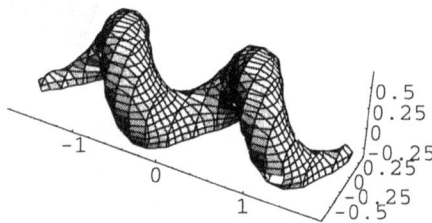

Fig 7. Putting lids on gives the worm.

Of the screw we make a worm (fig 7) by putting lids on:

$$10^{0.5(y^2+z^2)} + 10^{z\cos 4x - y\sin 4x} +$$
$$10^{0.15} + 10^{-0.15} = 4 \tag{7}$$

Next is to make real screws using the cone equations and the helicoid:

$$10^{-0.2x^2+y^2+z^2} + 10^{z\cos 3x - y\sin 3x} = 10 \tag{8}$$

This is really a catenoidic helicoid - we have chopped off half of it in fig 8a. We think this is a real beauty so we show three more projections in 8b, c, and d.

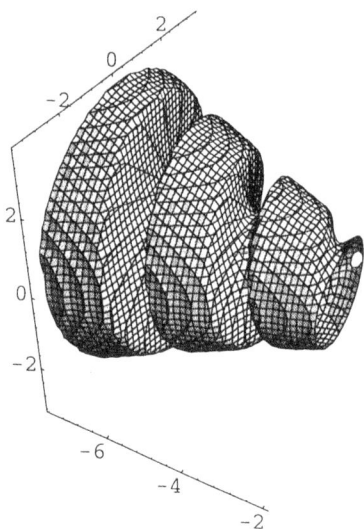

Fig 8a. Adding a cone gives a chiral funnel.

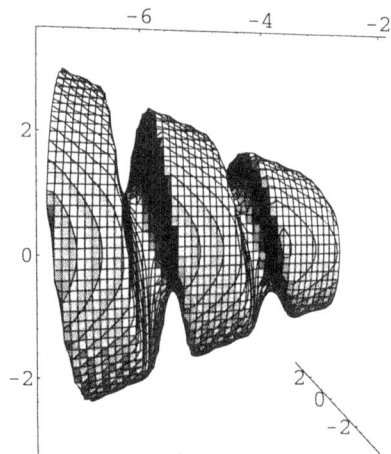

Fig 8b. Different view of **a**.

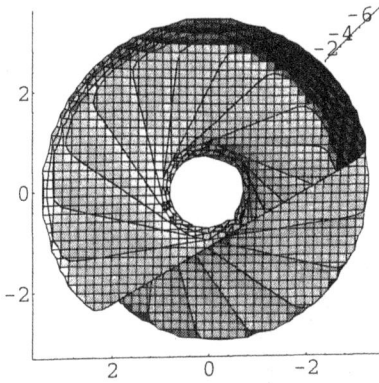

Fig 8c. An other view of **a**.   **Fig 8d.** A fourth view of **a**.

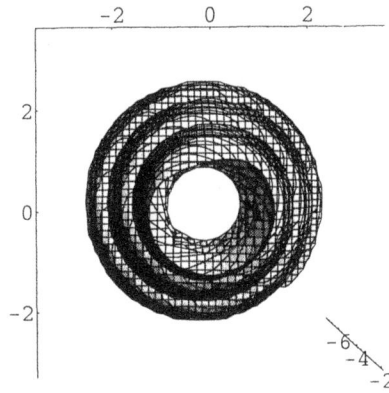

An obvious application for this screw is as funnel when a goose
is having his corn to eat.

We do the trick from equation (2) again and close the cone:

$$10^{0.2(-4x+y^2+z^2)} + 10^{z\cos 2x - y\sin 2x} = 10000 \qquad (9)$$

And we get a remarkable nose cone screw in fig 9a.

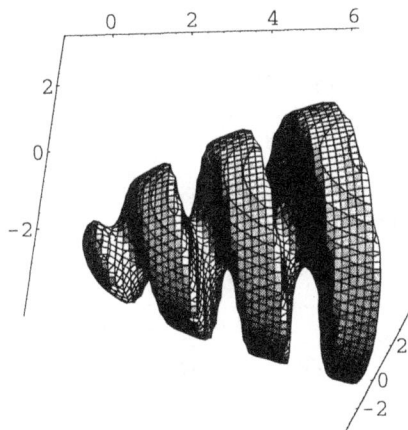

Fig 9a. Closing the cone gives the wooden
screw.

Before ending this session of screws we must of course give the ordinary screw. The equation is

$$e^{0.1(y^2+z^2)} + e^{z\cos 3x - y\sin 3x} = 10 \tag{10}$$

and the plot is in fig 9b.

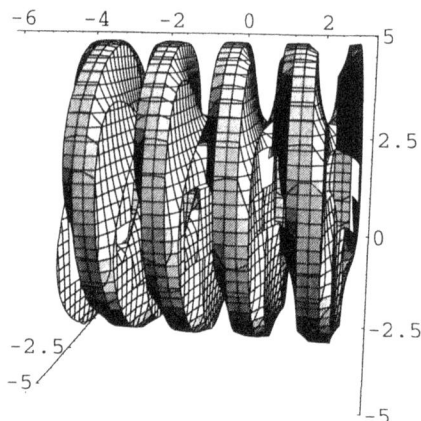

Fig 9b. The machine screw.

A very remarkable molecule is ferritin, an enzyme with a cavity containing up to 4500 iron atoms in form of a small inorganic crystal of the mineral ferrihydrite. This iron storage molecule is roughly spherical with rhombic dodecahedral symmetry, which gives access to the iron via six fourfold axis channels, and eight three fold. In chapter 5 we derived surfaces and nets for this geometry. In order to arrive at a shape for this molecule we use equation (15) and add a sphere.

$$10^{\cos 2\pi(x+y)} + 10^{\cos 2\pi(x-y)} + 10^{\cos 2\pi(y+z)} +$$
$$10^{\cos 2\pi(y-z)} + 10^{\cos 2\pi(x+z)} + 10^{\cos 2\pi(x-z)} + \tag{11}$$
$$10^{2(x^2+y^2+z^2)} = 20$$

This function has indeed a molecular shape as seen in fig 10a.
The four fold resp three fold channels are visualised in b and c.
Split versions are in d and e.

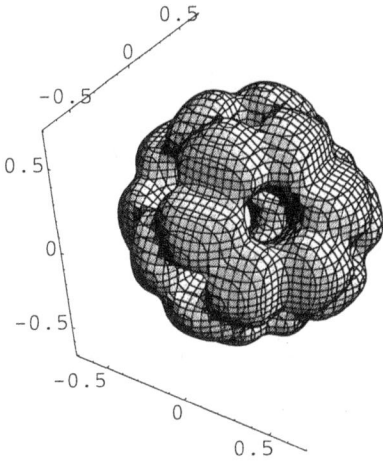

Fig 10a. Eq. (11) and the enzyme ferritine.

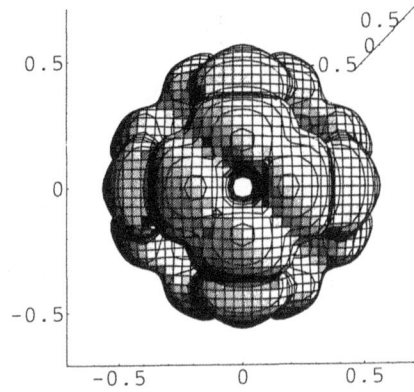

Fig 10b. The four fold channels, or entrances, to the giant molecule.

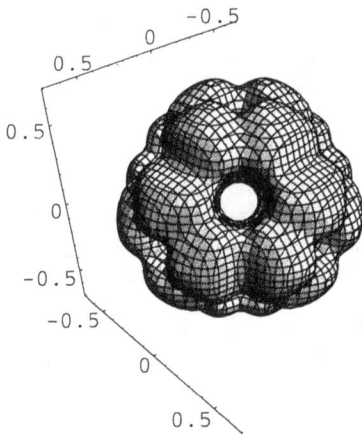

Fig 10c. The three fold channels.

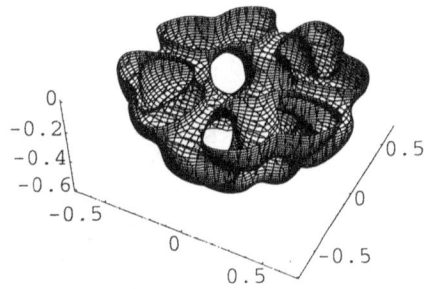

Fig 10d. A split of **a**.

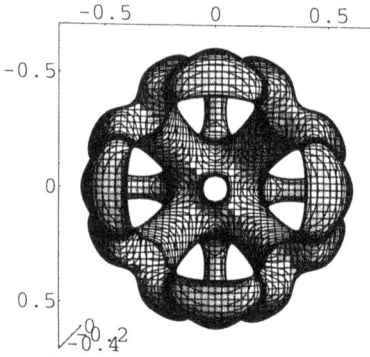

Fig 10e. A split along the four fold axis.

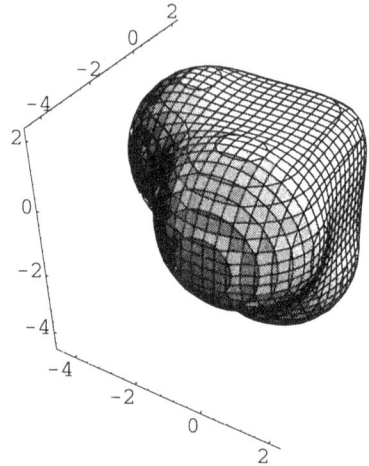

Fig 11. Spherical closing of a monkey saddle gives a cube corner closed with three spheres.

In a recent article[20] we used this technique to describe cubosomes, which are lipid bilayer particles with structures related to the minimal surfaces as discovered by Larsson[22,17]. A very simple example is the closing of the famous monkey saddle surface.

$$10^x + 10^y + 10^z - 10^{-x} - 10^{-y} - 10^{-z} +$$
$$10^{0.22(x^2+y^2+z^2)} = 100 \qquad (12)$$

This is perhaps not really a cubosome but the result is remarkable - a fusion of a cub corner and three spheres (fig. 11). Using a periodic surface as Larsson originally proposed gives beautiful cubosome structures. So we do the following calculation:

$$10^{(\cos 2\pi x \cdot \cos 2\pi y \cdot \cos 2\pi z + \sin 2\pi x \cdot \sin 2\pi y \cdot \sin 2\pi z)} +$$
$$10^{x^2+y^2+z^2} - 4 = 0 \qquad (13)$$

In fig 12a we see the cubosome, and in b we have opened it up to see that the D- surface is really in the inside.

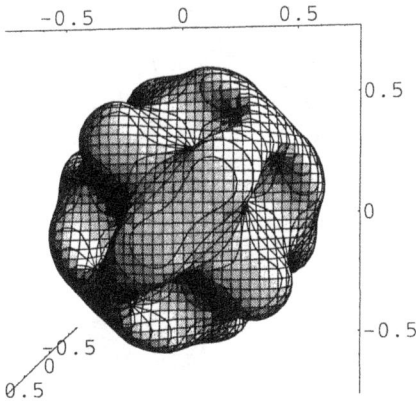

Fig 12a. Spherical closing of the D-surface gives a Larsson cubosome.

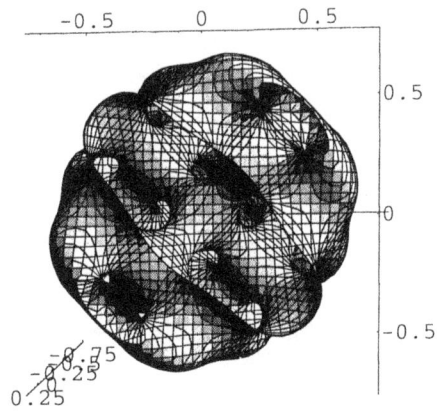

Fig 12b. A split to show the inside.

Before ending this book we will show you some more and remarkable surfaces. We start subtracting a cylinder from a cube after equation (14).

$$10^x + 10^y + 10^z + 10^{-x} + 10^{-y} + 10^{-z} -$$
$$10^{0.2(x^2+y^2)} = 370 \tag{14}$$

or, with the short hand notation,

$$10^{cube} - 10^{0.2cyl} = 370$$

The result is shown in three figures, 13a, b, and c, and speaks for itself.

We have done the same with the tetrahedron using the equation:

$$10^{tetra} - 10^{2cyl} = const \tag{15}$$

The equation for three constants, 100, 7 and 0, are respectively shown in figs 14a, b, c and d.

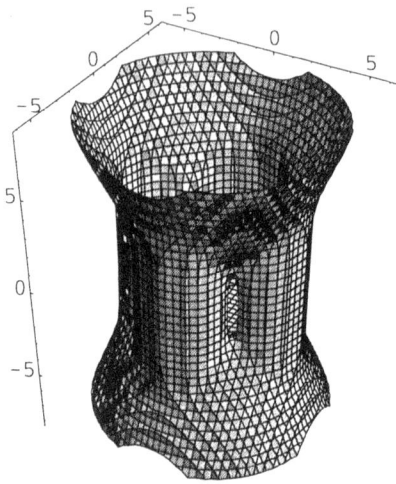

Fig 13a. A cylinder is subtracted from a cube.

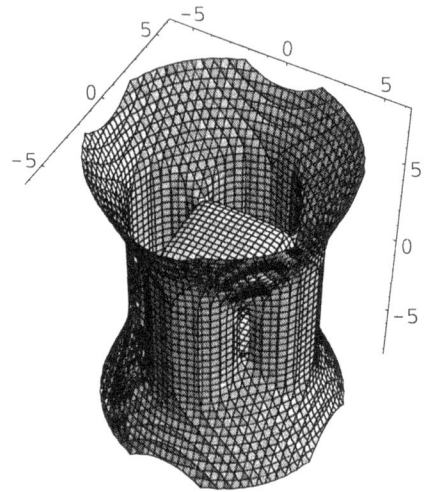

Fig 13b. Different view of **a**.

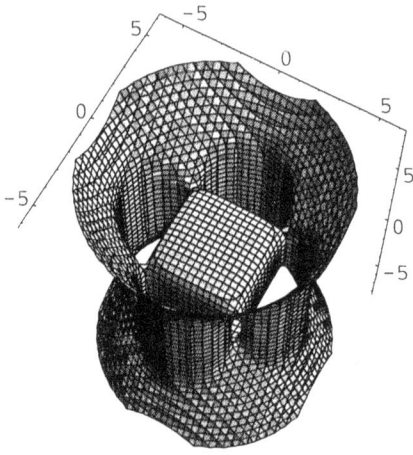

Fig 13c. A third view.

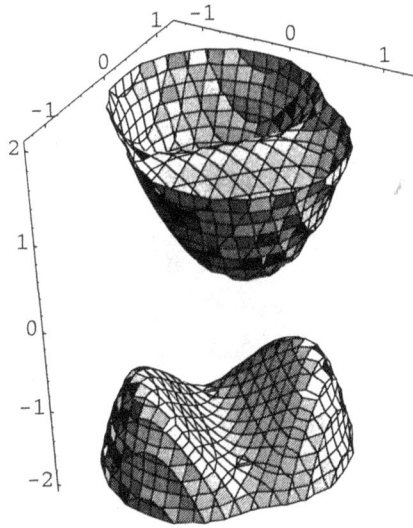

Fig 14a. A cylinder is subtracted from a tetrahedron with eq. (15) and the constant 100.0.

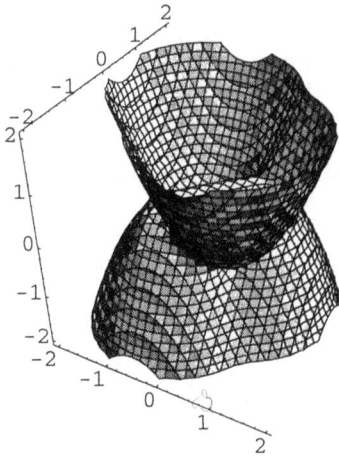

Fig 14b. Same as **a**, but const=7.0.

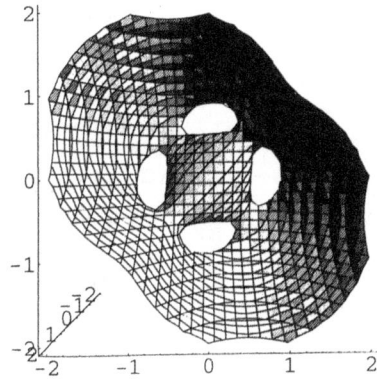

Fig 14c. Different view of **b**.

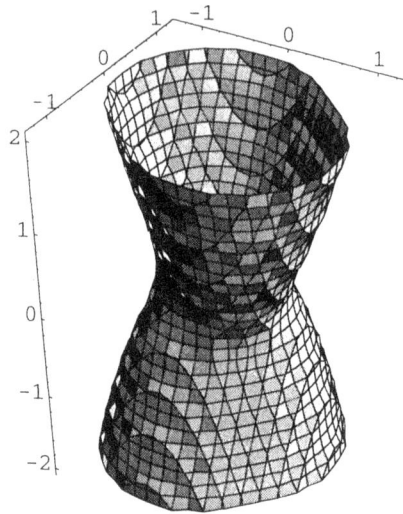

Fig 14d. Same as **a**, but const=0.0.

At a constant of 100 we see two saddle tipped cones approaching so that they at 7 fuse via a tetrahedron, as shown in fig 15a and b. At a constant of 0 the tetrahedron is dissolved into a catenoid of tetrahedral symmetry.

We have done the same with the octahedron and the icosahedron.

We can put spheres into a cylinder by using the P- surface, as seen in equation (16) and fig. 15a, b, c, and d.

$$10^{(\cos 2\pi x + \cos 2\pi y + \cos 2\pi z)} + 10^{4(x^2+y^2)} = \text{const} \qquad (16)$$

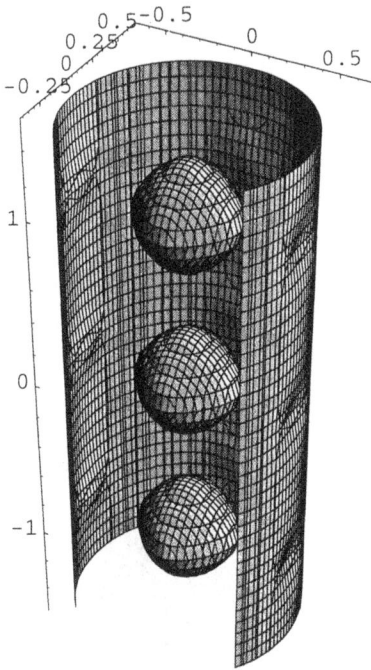

Fig 15a. Eq. (16) gives spheres in a cylinder with the constant 50.0.

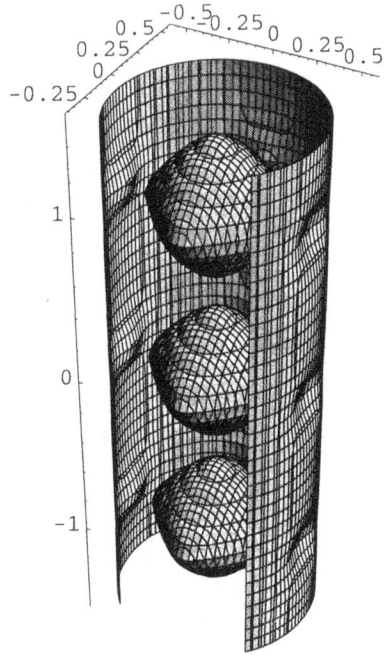

Fig 15b. Same as **a**, but with the constant 20.0.

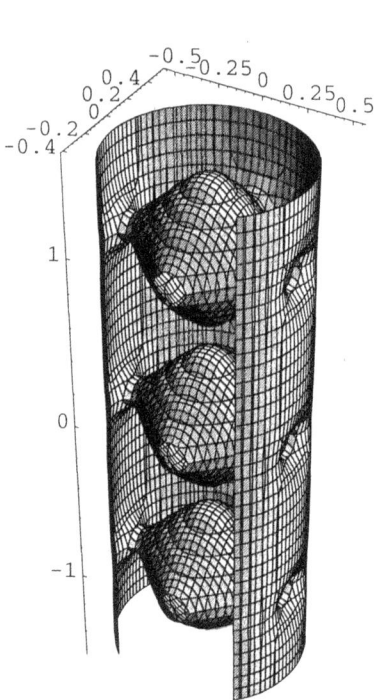

Fig 15c. Const = 15.0.

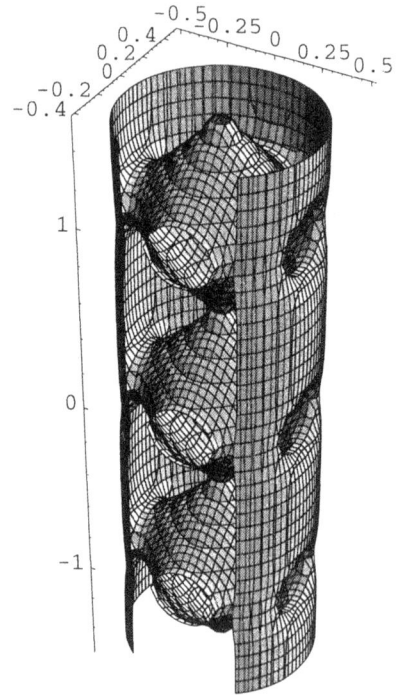

Fig 15d. Const = 10.0 making a row of octahedra.

The constant is varied down to 10 and we see how the spheres interact with each other and the cylinder to form a part of the P-surface, or a string of octahedra.

There are two classic surfaces in differential geometry - the catenoid and the helicoid- which are related by the Bonnet

transformation, as minimal surfaces they are. Here we simply
add them together and we see in fig 16 that a chiral catenoid is
the result. The equation is:

$$x\cos(\frac{\pi}{2}z) - y\sin(\frac{\pi}{2}z) + x^2 + y^2 - e^z - e^{-z} = 0 \quad (17)$$

For this remarkable topology the 'Gedankespiel' is that we have
a sieve for chiral molecules, which would pick up rotation
passing through. Making it easy for them to pass some other
chiral enzyme. Do the so called barrel enzymes have this
topology?

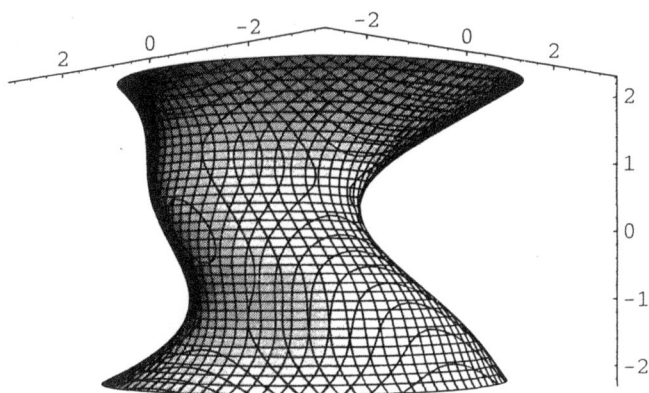

Fig 16. Chiral catenoid.

ACKNOWLEDGEMENTS

We wish to thank

- The Swedish Natural Research Council for financial support.
- Juri Grin at the Max Planck Institute in Stuttgart, and Reinhard Nesper at the ETH in Zürich for helpful discussions.
- Staffan Lindström at Studentlitteratur in Lund for his expertise and great kindness.

APPENDIX 1

CURVATURE CALCULATION

The following is a Mathematica Notebook code for calculation of
the Gaussian and mean curvature of a function w[x,y,z]. It is
written by Stephen Hyde, Applied Mathematics, ANU,
Canberra, Australia, and works for the program Mathematica.

```
w1[x_,y_,z_]    := D[w[x,y,z],x];
w2[x_,y_,z_]    := D[w[x,y,z],y];
w3[x_,y_,z_]    := D[w[x,y,z],z];
w11[x_,y_,z_]   := D[w1[x,y,z],x];
w12[x_,y_,z_]   := D[w1[x,y,z],y];
w13[x_,y_,z_]   := D[w1[x,y,z],z];
w21[x_,y_,z_]   := D[w2[x,y,z],x];
w22[x_,y_,z_]   := D[w2[x,y,z],y];
w23[x_,y_,z_]   := D[w2[x,y,z],z];
w31[x_,y_,z_]   := D[w3[x,y,z],x];
w32[x_,y_,z_]   := D[w3[x,y,z],y];
w33[x_,y_,z_]   := D[w3[x,y,z],z];

matrix[w_,x_,y_,z_]:={
    {w11[x,y,z]-l, w12[x,y,z],   w13[x,y,z],   w1[x,y,z]},
    {w21[x,y,z],   w22[x,y,z]-l, w23[x,y,z],   w2[x,y,z]},
    {w31[x,y,z],   w32[x,y,z],   w33[x,y,z]-l, w3[x,y,z]},
    {w1[x,y,z],    w2[x,y,z],    w3[x,y,z],    0}}

det[w_,x_,y_,z_]:=Det[matrix[w,x,y,z]];

a[w_,x_,y_,z_] := Coefficient[det[w,x,y,z],l,2];
b[w_,x_,y_,z_] := Coefficient[det[w,x,y,z],l,1];
c[w_,x_,y_,z_] := Coefficient[det[w,x,y,z],l,0];

meancurv[w_,x_,y_,z_]:=-b[w,x,y,z]/
(2*a[w,x,y,z]*Sqrt[w1[x,y,z]^2+w2[x,y,z]^2+w3[x,y,z]^2]);

gausscurv[w_,x_,y_,z_]:=c[w,x,y,z]/
(a[w,x,y,z]*(w1[x,y,z]^2+w2[x,y,z]^2+w3[x,y,z]^2));
```

REFERENCES

1   S.Andersson and B.G. Hyde,
     Z. Kristallogr. **158**, 119 (1982).

2   B.G.Hyde and S.Andersson,
     Inorganic Crystal Structures,
     Wiley, New York, 1988.

3   S.Hildebrandt and A. Troma,
     Mathematics and optimal form,
     Scientific American Library, W.H. Freeman,New York, 1985.

4   R. Mc.Lachlan,
     The Mathematical Intelligencer, **16**, 31 (1994).

5   W.P. Thurston,
     Scientific American, **251**, 94 (1984).

6   S.Lidin, S. Andersson, A. Fogden, M. Jacob and Z. Blum,
     Aust. J. Chem. **45**, 1519 (1992).

8   M. Jacob,
     Z. Kristallogr. **209**, 925 (1994).

9   S. Lidin, M. Jacob and S. Andersson,
     J. Solid State Chem. **114**, 36 (1995).

10   S. Andersson, S.T. Hyde and H. G. von Schnering,
     Z. Kristallogr. **168**, 1 (1984).

11   S. Andersson, S.T. Hyde, K. Larsson and S. Lidin,
     Chemical Reviews, **88**, 221 (1988).

12   S.T. Hyde and S. Andersson,
     Z. Kristallogr. **174**, 225 (1986).

13   S.T. Hyde, S. Andersson, B.Ericssonand and K. Larsson,
     Z. Kristallogr. **168**, 213 (1984).

14   H.G. von Schnering and R. Nesper,
     Angew.Chem. Int. Ed. Engl. **26**,1059 (1987).

15   H.G. von Schnering and R. Nesper,
      Z. Phys. B - Condensed Matter **83**, 407 (1991).

16   U. Dierkes, S. Hildebrandt, A. Kuster and O. Wohlrab,
      Minimal Surfaces 1 and 2, Springer Verlag, Berlin, 1991.

17   K. Larsson,
      *Lipids - Molecular Organization, Physical Functions and
      Technical Applications,* The Oily Press, Dundee, 1994.

18   S. Andersson, M. Jacob and S. Lidin,
      Z. Kristallogr. **210**, 3 (1995).

19   S. Andersson, M. Jacob and S. Lidin,
      Z. Kristallogr. **210**, 826 (1995).

20   S. Andersson, M. Jacob, K. Larsson and S. Lidin,
      Z. Kristallogr. **210**, 315 (1995).

21   A. Burkhardt, U. Wedig and H. G. von Schnering,
      Z. anorg. allg. Chem. **619**, 437 (1993).

22   K. Larsson,
      J. Phys. Chem. **93**, 7304 (1989).

23   H. Schäfer, H.G. von Schnering, K.J. Nieves and H.G.
      Nieder-Vahrenholz,
      J. Less Common Metals **9**, 95 (1965).